AI

The Future Unveiled

By
Alex Trenton

AI

The Future Unveiled

Table of Contents

Introduction

Welcome to an eye-opening journey into the world of Artificial Intelligence. If you're holding this book, you're probably someone who's intrigued by how AI is transforming industries, creating new possibilities, and challenging our traditional ways of thinking. Whether you're a tech enthusiast eager to dive into the specifics, a business leader looking to leverage AI for competitive advantage, or simply someone curious about what the future holds, this book aims to provide you with the knowledge and strategies you need to navigate the AI-driven future.

AI isn't just a buzzword; it's a paradigm shift. Every day, we're witnessing advancements that seemed like science fiction merely a few decades ago. From self-driving cars to intelligent personal assistants, AI is becoming an integral part of our daily lives. And while these innovations bring exciting opportunities, they also raise essential questions about ethics, privacy, and the future of work. This book strives to answer those questions, offering insights into the multi-faceted world of AI.

Yet, the topic of AI isn't just about technology. It's about understanding who we are and who we could become. AI challenges our notions of intelligence, productivity, and even what it means to be human. The stakes are high, and the implications are profound. Thus, it's crucial that we approach this topic with both curiosity and caution, armed with the knowledge to make informed decisions as we move forward.

You'll find that this book is organized into chapters that meticulously cover a broad spectrum of AI applications and impacts. In the first few chapters, we'll set the stage by defining AI and tracing its historical development. You'll learn about different types of AI, the initial experiments, and the seminal breakthroughs that have shaped where we are today. This foundational understanding is crucial as we delve deeper into more complex topics.

But we won't stop at just the historical perspective. The book then catapults into the present and future, exploring emerging trends such as machine learning and deep learning. We'll discuss how AI is revolutionizing sectors like healthcare, economics, and transportation, making our systems more efficient and our lives more personalized. The goal is to give you a panoramic view of AI's transformative potential, illustrating both the opportunities and challenges that lie ahead.

For instance, in healthcare, AI-driven diagnostic tools and personalized medicine are redefining patient care. In economics, predictive analytics are reshaping market trends and financial forecasting. These aren't just abstract concepts; they are real-world applications that are making a tangible difference.

One of the core aims of this book is to demystify AI. Often, complex jargon and technical details can make AI seem inaccessible. We'll break down these barriers, offering clear explanations and real-life examples that make the subject matter relatable and easy to grasp. This isn't a book written solely for tech wizards. It's for anyone who wants to understand the profound changes AI brings to various facets of our lives.

Moreover, we'll delve into the ethical considerations and societal impacts of AI. What does AI mean for data security and personal privacy? How do we ensure that AI development is fair and unbiased? These are questions without simple answers, but they are questions we must grapple with. Our discussions will aim to shed light on these complex issues, providing multiple perspectives and compelling arguments.

In the business realm, AI is a game-changer. Whether it's optimizing operational efficiency, enhancing customer experiences, or creating new business models, AI offers unprecedented opportunities for innovation and growth. We'll look at case studies and industry implementations to understand how businesses are successfully leveraging AI.

But there's another side to the coin: the impact of AI on the workforce. While AI has the potential to displace jobs, it can also create new opportunities and roles that we haven't yet imagined. Understanding this dichotomy is crucial for strategic planning and skill development.

We'll also explore how AI is revolutionizing other critical areas like communication, entertainment, and education. From natural language processing to adaptive learning systems, AI is making interactions more intuitive and learning more personalized. These advancements promise to create a more connected and informed society.

On the environmental front, AI is playing a pivotal role in climate modeling and resource management. By harnessing the power of AI, we can better understand and address the pressing issues of environmental sustainability, paving the way for a greener future.

Security is another critical area where AI is making significant inroads. Cybersecurity threats are becoming increasingly sophisticated, and traditional defenses are often inadequate. We'll explore how AI can bolster our security measures, from cybersecurity to national defense.

The societal impacts of AI are far-reaching. Issues of social inequality and public perception are critical as AI technologies become more integrated into our daily lives. We must consider who benefits from AI and who doesn't, ensuring that the advantages of AI don't exacerbate existing inequalities.

Innovation is at the heart of AI's potential. This book will cover how AI is driving research and development, as well as its role in patent analysis and intellectual property. We'll look at how AI is fostering a new wave of creativity and problem-solving across various industries.

As we examine real-world applications of AI, you'll see how different sectors are implementing these technologies to solve specific challenges. These case studies offer valuable lessons and inspire new ideas, illustrating the tangible benefits AI can bring.

Of course, the journey of AI is fraught with ethical dilemmas. Issues of accountability, bias, and fairness are at the forefront of AI development. We'll explore these concerns in depth, offering perspectives from experts in the field and presenting frameworks for ethical AI development.

The regulatory landscape for AI is evolving, with existing laws and future policy considerations shaping how AI technologies are developed and deployed. We'll provide an overview of the regulatory environment, highlighting key legal frameworks and ongoing debates.

Investing in AI is another area of interest for many readers. Whether you're an investor looking to capitalize on AI trends or a startup seeking venture capital, understanding the market dynamics is crucial. We'll offer insights into market analysis, investment opportunities, and the venture capital landscape.

This book also focuses on preparing for the AI revolution. Skill development and strategic planning are essential for anyone looking to thrive in an AI-driven world. We'll discuss the skills needed to stay relevant and strategies for harnessing the power of AI for personal and professional growth.

In the latter chapters, expert opinions will provide predictive analysis and a global perspective on AI. From interviews with leading thinkers to insights into worldwide adoption and international collaboration, these sections offer a comprehensive view of AI's global impact.

Finally, as we look toward the future, we'll explore AI's role in emerging technologies like quantum computing and advanced robotics. These advancements promise to push the boundaries of what's possible, opening up new horizons and potential challenges.

The road ahead for AI is filled with potential and uncertainty. By staying informed and engaged, we can better navigate the evolving landscape and harness AI's power responsibly and effectively.

In conclusion, this book aims to be your comprehensive guide to understanding and engaging with the world of AI. So, let's embark on this journey together, exploring the depths of AI, its impact on our world

Chapter 1:
Understanding AI

Jumping right into the world of artificial intelligence (AI), we find ourselves on the precipice of a revolution that promises to reshape every facet of our lives. AI isn't just a piece of jargon tossed around in tech circles; it's a transformative force driving innovation across industries, from healthcare to transportation, and even entertainment. It's essential to grasp what AI truly is—a complex amalgamation of algorithms, data, and machine learning that mimics human cognitive functions such as learning and problem-solving. But AI isn't one-size-fits-all. There are various types, each with distinct capabilities and applications, ranging from narrow AI, which excels in specific tasks, to the conceptual realms of general and superintelligent AI. Understanding AI means unlocking the potential to leverage it responsibly, pushing our boundaries while ensuring ethical considerations aren't left in the dust. This chapter sets the stage for a deeper dive into the nuances of AI, shedding light on its types, and preparing you for the profound impact it will have on our future.

What is AI?

Artificial Intelligence, or AI, is the concept of building machines that can think, act, and learn in ways that mimic human intelligence. At its core, AI straddles the intersection of technology and cognition. Imagine a computer that can make decisions, learn from experience, and even adapt to new situations without human intervention. It's both a marvel and a mystery—a blend of algorithmic wizardry and logical constructs that it's changing the way we interact with the world and how the world interacts with us.

Defining AI isn't exactly straightforward, though. It's something that resists a single, concise definition because it spans such a broad scope of functionalities and applications. AI can range from simple rule-based systems to sophisticated neural networks that approximate the human brain's structure. The essence of AI lies in creating systems capable of performing tasks that would typically require human intelligence—think problem-solving, recognizing patterns, understanding natural language, and even exhibiting emotional responses.

These systems are powered by data—lots of it. In fact, AI thrives on data. The more information it processes, the more intelligent it becomes. Think of an AI as an ever-evolving entity, continuously learning and improving from every bit of data it ingests. This continuous learning loop is what makes AI so potent and transformative. And while we're still scratching the surface of what AI can do, the potential seems almost limitless.

However, AI isn't a monolithic technology; it's more of an umbrella term encompassing various subfields and methodologies. At its core, AI can be broken down into two primary types: Narrow AI and General AI. Narrow AI, also known as Weak AI, is designed to perform a narrow task—such as facial recognition or internet searches. On the other hand, General AI (or Strong AI) aims to outperform humans at nearly every cognitive task.

Take Siri, for example. Apple's voice assistant can handle a plethora of tasks, from setting reminders to providing weather updates, making it a standout example of Narrow AI. It excels in specific areas but can't generalize its learning to other domains outside its programmed capabilities. In contrast, General AI remains largely theoretical at this point. This is the kind of AI that dominates science fiction, capable of understanding, learning, and applying intelligence across a multitude of contexts just like a human being. We're still a long way from achieving it, but researchers are making incremental strides every day.

Another way to understand AI is through its mechanisms. Machine Learning (ML) and Deep Learning (DL) are the engines driving most of today's AI advancements. Machine Learning involves training algorithms to find patterns and make decisions based on data. It's what helps Netflix recommend your next binge-watch or what allows self-driving cars to understand road signs and avoid obstacles. Deep Learning, a subset of ML, uses artificial neural networks to model complex patterns in data. These neural networks are designed to mimic the human brain, allowing for exceptionally high levels of abstraction and predictive power.

But AI isn't just about algorithms and data; it's about solving real-world problems. AI can diagnose diseases, predict stock market trends, customize our online shopping experiences, and even compose music. It holds the potential to revolutionize virtually every industry, creating new business models and transforming existing processes. For businesses, embracing AI could be the difference between staying competitive and falling behind. The age-old mantra, "adapt or perish," has never been more relevant.

While the technological underpinnings of AI are undeniably complex, the impact of AI is something anyone can grasp. It's here to make life easier, more efficient, and in many cases, more personalized. Imagine a world where your digital assistant understands your needs before you even articulate them, or where medical diagnoses are not only faster but also more accurate. That's the world AI is steering us towards, a future where intelligence isn't a trait confined to human beings but is embedded into the very fabric of our daily lives.

Understanding AI also requires grappling with its ethical and social implications. The inclusion of AI in decision-making processes raises fundamental questions about privacy, accountability, and fairness. For instance, how do we ensure that AI systems do not propagate existing biases or create new ones? What safeguards do we install to prevent the misuse of such powerful technology? Navigating these ethical dilemmas is just as crucial as technological advancements themselves. As AI continues to evolve, so must our frameworks for ethical considerations and governance.

AI is not without its challenges. Data quality, interpretability, and scalability are significant hurdles. Even the brightest minds in the field wrestle with these issues, striving to make AI systems not only smarter but also more transparent and equitable. The best AI systems are those that can explain their reasoning, enabling users to trust their results and understand the underlying processes. In a world increasingly driven by data, transparency and trust are paramount.

We are currently in what many call the golden era of AI. Funding is abundant, research is thriving, and the pace of innovation is staggering. New algorithms and paradigms emerge almost weekly, challenging our previous assumptions and pushing the boundaries of what's possible. Yet, for all its rapid advancements, AI still requires human collaboration. It's a partnership—a synergistic dance where human ingenuity guides, and machine precision executes.

AI's journey is reminiscent of the early days of computing. When computers first emerged, they were seen as colossal calculators, their potential applications barely imaginable. Today, it's almost unthinkable to live without them. Similarly, while AI may seem daunting or esoteric now, it's poised to become an integral part of our everyday lives. In the future, saying you don't use AI might sound as odd as saying you don't use electricity.

It's exciting to think about the possibilities. AI's potential applications span education, healthcare, finance, entertainment, transportation—the list goes on. It's not just about automating tasks; it's about augmenting human capabilities. It's about collaboration, not just replacement. The future of AI isn't just about smarter machines; it's about smarter living, smarter decision-making, and ultimately, a smarter world.

As we venture into this AI-driven future, it's crucial to remain both optimistic and vigilant. AI brings promise but also responsibility. Embracing it means preparing adequately—understanding its potentials, its limits, and its ethical dimensions. And as you delve deeper into this book, you'll gain insights into how to navigate this brave new world, armed with the knowledge to harness AI's full potential while mitigating its risks. Welcome to the fascinating world of Artificial Intelligence!

Types of AI

Understanding the types of AI is crucial for anyone keen on navigating the complexities of this fascinating field. Broadly speaking, AI can be classified into three categories: Narrow AI, General AI, and Superintelligent AI. Each type operates at a different level of capability and has distinct potential applications and implications.

Narrow AI: This is the most prevalent type of AI that we encounter today. Also known as Weak AI, Narrow AI is designed and trained for a specific task. Examples are all around us. Think of Siri, Alexa, or even Google's search algorithms. These systems can handle narrowly defined tasks remarkably well but struggle outside their specified domain. For instance, a chess-playing AI might be unbeatable at chess but can't diagnose a medical condition.

Narrow AI excels in specialized functions, thanks to advancements in machine learning and deep learning. For instance, chatbots that handle customer service inquiries or AI systems used in recommendation engines on streaming services. The efficiency and reliability of Narrow AI systems have made them indispensable in various industries, enhancing productivity and improving user experiences.

Interestingly, many industries leverage Narrow AI to drive innovation. Manufacturing uses it to predict equipment failures, avoiding costly downtimes. Healthcare employs Narrow AI for reading radiographs more accurately than human radiologists. In finance, it's all about algorithmic trading where AI models analyze market data at lightning speed to make trades in milliseconds. With its high specialization, Narrow AI is both a boon and a limitation; it's powerful within its scope but can't generalize beyond it.

General AI: Often referred to as Strong AI or AGI (Artificial General Intelligence), this type aims to possess the same level of intelligence and understanding as a human being. It can perform any intellectual task that a human can, adapting to different situations and learning from experiences. While still a theoretical concept, General AI signifies performing complex activities like reasoning, problem-solving, and abstract thinking across various domains without human intervention.

Imagine an AI doctor that not only diagnoses diseases accurately but also empathizes with patients, or an AI attorney that understands and interprets legal nuances as well as any human lawyer. The development of General AI would revolutionize every sector, rendering it a universal problem solver. The fascinating aspect of General AI lies in its potential to understand human emotions, develop social skills, and even engage in philosophical discussions. It's like bringing human-like wisdom and versatility into machine form.

However, achieving General AI involves staggering challenges. Developing algorithms that can reason, form strategies, and understand context in a manner akin to human cognition is a massive undertaking. Ethical considerations and safety are paramount, given the immense power such systems would hold. Researchers are exploring a range of approaches, from cognitive architectures to whole brain emulation, yet most projections suggest it may take decades before we reach true General AI.

Superintelligent AI: Superintelligent AI goes beyond both Narrow and General AI. This type refers to an AI that surpasses human intelligence in every imaginable field—surpassing not just logical or mathematical tasks but also in social benchmarks. Superintelligent AI could conceive solutions to problems that human intellect might never grasp. It raises questions about what it means to be intelligent and how we measure and understand cognition.

Superintelligent AI introduces fascinating possibilities—curing diseases currently deemed incurable, solving climate change, even advancing technologies at an unprecedented rate. But with great power comes great responsibility. The advancement of such an all-encompassing intelligence has prompted debates about existential risks. How do we control something vastly more intelligent than ourselves? Ensuring Superintelligent AI aligns with human values and goals is an ongoing ethical dilemma.

In navigating these types, it's essential to understand where we stand today and what the future might hold. Narrow AI is already revolutionizing industries; it's the here and now. General AI represents an ambitious frontier that could reshape our world and our understanding of intelligence. And Superintelligent AI offers the most transformative potential but with challenges that touch on our very existence.

Work towards these AI types continues in labs and think tanks worldwide. The symbiosis of machine learning algorithms, ever-growing data sets, and advancements in processing power pushes us closer to realizing the dreams—and addressing the fears—these AI types inspire. For tech enthusiasts and business leaders, keeping an eye on these developments isn't just about staying relevant; it's about gaining the foresight necessary to adapt and thrive in an AI-driven future.

Chapter 2:
The History of AI

Starting from speculative musings in ancient myths to the sophisticated neural networks powering today's innovations, the history of AI is a captivating journey through human ingenuity and curiosity. It began with early attempts in the 1950s to create "thinking machines" that could mimic human reasoning. As time marched forward, pivotal milestones such as the establishment of the Dartmouth Conference in 1956 laid the groundwork, sparking an era of rapid development. The narrative of AI's history is dotted with breakthroughs like IBM's Deep Blue defeating chess grandmaster Garry Kasparov in 1997 and more recently, the advent of deep learning technologies that can recognize and generate human-like text, images, and even voices. Each advancement, backed by relentless research and evolving computational power, nudges us closer to a future where the line between human and machine intelligence becomes increasingly blurred. This rich and evolving tapestry of AI has not only reshaped industries but also continues to redefine the essence of what it means to be intelligent in the modern world.

Early Beginnings

Looking back to the early beginnings of AI, it might feel like diving deep into the pages of a fascinating story. AI didn't spontaneously spring into existence; it was born from the need to understand and replicate human intelligence. The seeds of this journey were planted long before computers became household items, and the early chess moves of AI development were surprisingly strategic and ambitious.

It's tempting to think of AI as a 21st-century marvel, but people have been curious about artificial intelligence for centuries. The concept of machines that could think and learn dates back to ancient myths. Philosophers like Aristotle mused over mechanical beings that could mimic human capabilities. These philosophical inquiries laid the foundation for the evolution of AI, driving us to uncover the mysteries of cognition.

The story of AI formally begins in the mid-20th century, when the convergence of philosophy, mathematics, and engineering gave birth to the first serious attempts at creating a thinking machine. In 1950, Alan Turing laid out his famous "Turing Test," a criterion for determining whether a machine could exhibit intelligent behaviour indistinguishable from a human. This was a bold statement that dared to imagine machines as intellectual equals to men.

Turing's groundbreaking work didn't occur in isolation. The following years saw a flurry of activity from brilliant minds. In 1956, the term "Artificial Intelligence" was coined at the now-famous Dartmouth Conference, organized by John McCarthy, Marvin Minsky, Nathaniel Rochester, and Claude Shannon. It was here that the vision for AI as a dedicated field of study took root. The attendees were a mix of computer scientists, linguists, and engineers who shared a common goal: to explore the possibilities of creating machines that could learn and reason like humans.

The enthusiasm generated by the Dartmouth Conference was electric, sparking a rush of funding and research. Early pioneers like McCarthy and Minsky began developing the foundational theories and algorithms that would fuel subsequent advancements. They focused heavily on problem-solving and symbolic methods, which dominated the field for decades.

One exemplary effort was the Logic Theorist, developed by Allen Newell and Herbert A. Simon in the late 1950s. This program was designed to mimic human problem-solving skills, specifically in proving mathematical theorems. It was a success, and more importantly, it demonstrated that machines could carry out tasks that required human-like thought processes.

The 1960s and 1970s were periods of great optimism for AI. Researchers created programs that could play chess, solve algebra problems, and even understand human language to a limited extent. The excitement was palpable, and the potential seemed boundless. However, it wasn't all smooth sailing. The limitations of early computer hardware and the complexity of scaling up these problem-solving capabilities soon became apparent. The initial exuberance was tempered by hard realities: computing power was limited, and the algorithms were not as robust as needed for broader applications.

Funding started to wane as expectations outpaced reality. During the 1970s and 1980s, AI entered what is often referred to as the "AI Winter," a period marked by reduced interest and investment. The progress was slow, and the high expectations set during the early days weren't met, leading to disillusionment. Yet, those years were far from a complete loss. Researchers used this time to refine their approaches, focusing on more specialized applications and incremental advancements.

One silver lining of the AI Winter was the birth of expert systems. These systems, such as MYCIN, were designed to replicate the decision-making abilities of human experts in specific domains. MYCIN, in particular, was a medical diagnostic system that could recommend treatments for bacterial infections. Though limited to a narrow range, expert systems demonstrated the feasibility and utility of AI in real-world applications. They provided the stepping stones from theoretical exploration to practical implementation.

While many may have written off AI as a failed experiment, the tireless work of researchers laid the groundwork for a resurgence that would come decades later. The lessons from these early years were invaluable, highlighting both the promise and the limitations of artificial intelligence. This period served as a testament to the enduring human spirit of curiosity and innovation.

Moreover, the early beginnings of AI were not just about algorithms and hardware. They were about the bold dreams that chased the question, "What if?" What if we could create machines that think and learn? What if we could unlock new levels of human potential through technology? These questions propelled a journey that continues to shape our present and future.

As we move forward in this book, tracing the trajectory of AI from its inception to its current state, it's crucial to remember these early days. They remind us that every groundbreaking technology starts with a spark of imagination. The early scientists and researchers dared to dream big, to venture into the unknown. They set in motion a series of events that would lead to the sophisticated AI systems we marvel at today.

The early chapters of AI's history are rich with stories of struggle, innovation, and resilience. Understanding these beginnings isn't just about appreciating the past; it's about recognizing the building blocks that support today's advancements. It's about drawing inspiration from the visionaries who dared to think differently, who believed in a future where machines could complement human intelligence.

So, as we delve deeper into the world of AI, let's keep in mind the giants on whose shoulders we stand. Their work teaches us that progress is a winding path, full of unexpected turns but always driven by the quest for knowledge. They remind us that today's challenges are just the launchpad for tomorrow's breakthroughs. The journey of AI is far from over, and the best is yet to come.

Key Milestones

The story of artificial intelligence (AI) began long before Siri answered your first question or your car parallel-parked itself. Understanding its key milestones helps us not only appreciate how far we've come but also envision the enormous possibilities that lie ahead.

One of the earliest milestones in AI history can be traced back to 1950 when Alan Turing, the British mathematician and logician, published his seminal paper "Computing Machinery and Intelligence." In it, he posed the provocative question, "Can machines think?" and introduced what later became known as the Turing Test. This concept evolved into a benchmark for determining a machine's ability to exhibit intelligent behavior equivalent to, or indistinguishable from, that of a human.

The 1956 Dartmouth Conference is often referred to as the birth of AI as a field. Organized by John McCarthy, Marvin Minsky, Nathaniel Rochester, and Claude Shannon, this conference brought together a group of researchers who were enthusiastic about the potential of machines to simulate any aspect of human intelligence. The term "artificial intelligence" was coined here, and it officially marked the beginning of AI as a recognized academic discipline.

Fast forward to the 1960s, and you find the creation of the first AI programs that could play games and solve problems. For instance, IBM's Arthur Samuel developed a checkers-playing program that learned from experience. Another pioneering effort was the "General Problem Solver" (GPS), created by Herbert A. Simon and Allen Newell. These initiatives demonstrated early forms of machine learning and problem-solving capabilities, seeding the path for future innovations.

A pivotal moment in AI came during the 1970s when the development of expert systems took center stage. These systems were designed to mimic the decision-making abilities of a human expert. One of the most famous was MYCIN, an expert system for diagnosing bacterial infections and recommending antibiotics. Though rudimentary by today's standards, MYCIN showcased the practical applications of AI in specialized fields, particularly in healthcare.

The arrival of backpropagation algorithms in the 1980s was another game-changing milestone. Backpropagation is essential for training neural networks, and its introduction opened up new avenues in the field of machine learning. This period also saw the development of commercial AI applications, particularly in manufacturing and logistics, where robots began to take on more complex tasks.

Then came the 1990s, a decade that marked significant leaps in computational power and algorithmic sophistication. IBM's Deep Blue became the first computer to beat a reigning world chess champion, Garry Kasparov, in 1997. It wasn't just a victory for machines over humans but a major breakthrough in demonstrating the capabilities of AI in strategic thinking and complex problem-solving. Interestingly, it was an achievement that both captivated and alarmed the world, solidifying AI's place in popular culture.

As we moved into the new millennium, advancements in natural language processing (NLP) began to flourish. In 2001, the release of iRobot's Roomba hinted at a future filled with smart, autonomous devices capable of assisting in daily chores. Five years later, the unveiling of IBM's Watson took the world by storm when it won the game show Jeopardy! in 2011 against human champions. Incorporating deep analytics and NLP, Watson demonstrated the next level of machine intelligence by understanding and processing human language in real-time.

Unarguably, the rise of deep learning in the 2010s has been one of the most transformative milestones in AI history. Technologies such as convolutional neural networks (CNNs) and generative adversarial networks (GANs) began to unlock previously unattainable possibilities in image recognition, speech synthesis, and more. Google's AlphaGo made headlines in 2016 by defeating the world champion Go player, a game famously more complex than chess. This achievement showcased AI's potential in navigating intricate tasks and making nuanced decisions, far beyond simpler game environments.

Another major AI milestone was the release of OpenAI's GPT-3 in 2020. With 175 billion parameters, GPT-3 achieved unprecedented results in natural language understanding and generation, propelling advancements in fields from automated content creation to customer service chatbots. It's not just a model but a glimpse into the future possibilities of AI-human interactions, offering a revolutionary toolset for businesses and creatives alike.

With exponential growth in AI's capabilities, the milestones are becoming more frequent and impactful. In healthcare, AI algorithms are now diagnosing diseases with an accuracy that rivals, and sometimes surpasses, human doctors. On the business front, AI-driven predictive analytics are transforming decision-making by providing deep insights into consumer behavior and market trends. Autonomous vehicles are no longer a futuristic concept; they're being tested on roads around the world, promising to revolutionize transportation.

What makes these milestones even more remarkable is their collective ability to solve real-world problems. They've not only pushed the boundaries of what's technologically possible but have also prompted important societal questions about ethics, data privacy, and the future of work. For every breakthrough, there's a corresponding challenge, and navigating these with wisdom and foresight will be crucial as we journey further along the AI trajectory.

Each milestone we've covered thus far hints at the limitless potential that lies ahead. AI's story isn't just about machines thinking or solving complex problems. It's about enriching human lives, democratizing technology, and fostering innovation across industries. As we continue to advance, we'll undoubtedly face obstacles, but history has shown that the AI community is resilient, driven by curiosity, and continually pushing toward new horizons.

When we look back at these milestones, we see a mosaic of human ingenuity and machine precision working in tandem. The interplay between visionaries like Turing, McCarthy, and modern-day researchers has created a dynamic field that constantly evolves and redefines the possible. The key milestones in AI's history serve as guideposts, showing us where we've been and hinting at where we're going.

As we anticipate future milestones, it's crucial to remember that every step forward in AI is a collective achievement—one that involves academia, industry, and society at large. In this grand adventure, the quest for artificial intelligence may very well lead us to a deeper understanding of our own intelligence and what it means to be human.

Chapter 3:
Emerging Trends in AI

The landscape of AI is developing at a breakneck pace, revealing groundbreaking trends that are set to redefine how we interact with technology and solve complex problems. The emergence of powerful machine learning algorithms is empowering systems to learn from vast amounts of data, enabling them to make ever-more accurate predictions and decisions. Simultaneously, advances in deep learning—a subset of machine learning modeled on the human brain's neural networks—are accelerating breakthroughs in image and speech recognition, natural language processing, and even autonomous systems. These technologies aren't just buzzwords; they're fundamental shifts that promise to disrupt industries, enhance human capabilities, and unlock unprecedented potential. As we peer into the future, it's clear that understanding and harnessing these trends will be key to navigating the uncharted waters of the AI-driven world. From improving healthcare outcomes to creating smarter cities, the possibilities seem almost limitless, each new development building on the last, making the future of AI an exhilarating frontier.

Machine Learning

Machine Learning (ML) stands as one of the most prominent branches within Artificial Intelligence (AI) and has become a catalyst for innovation across multiple industries. This technology involves training algorithms to learn from and make predictions based on data.

Essentially, these algorithms identify patterns and make decisions with minimal human intervention.

Now, imagine teaching a child to recognize different breeds of dogs. Initially, you show them images of various breeds, describe the distinctive features, and repeat the process until they can identify the breeds independently. Machine Learning operates similarly, but on a much grander scale. With the right data and algorithms in place, a machine can, for example, distinguish between spam and legitimate emails or even diagnose medical conditions.

So, how exactly does ML work? At its core, the process involves feeding large sets of data into a model and using statistical methods to enable the model to make data-driven decisions. Algorithms such as decision trees, support vector machines, and neural networks serve as the backbone of this learning process. The end goal is for the system to improve its performance over time, adapting and refining its output based on continuous input.

Take supervised learning, for instance. In this approach, machines learn from labeled datasets, meaning each input comes with a known output. Think of it like a student learning with a teacher's guidance, where the answers are continuously checked and corrected until understood.

On the flip side, unsupervised learning deals with unlabeled data. Here, the machine's task is to identify hidden patterns or intrinsic structures within the input data. It's akin to solving a puzzle with no picture on the box to guide you.

One might ask, "Why all the hype about Machine Learning?" The answer lies in its transformative impact. Businesses leverage ML for predictive analytics to forecast market trends and optimize operational efficiency. In healthcare, it's being used for early disease detection, which can significantly improve patient outcomes.

However, it's essential to understand that ML isn't a magic wand. The effectiveness of a machine learning model hinges on the quality of data. Garbage in, garbage out, as the saying goes. Hence, data preprocessing, such as cleaning and normalization, becomes crucial steps in the ML pipeline.

Consider a sports team aiming to win a championship. They analyze opponents' strategies, refine their playbooks, and continuously practice until they outshine the competition. Similarly, ML models need to be tested, validated, and tuned rigorously to ensure optimal performance.

The rise of machine learning has also spurred advancements in other AI domains, particularly deep learning. With its layers of artificial neural networks modeled after the human brain, deep learning has been instrumental in achieving remarkable feats in image and speech recognition. But let's not wander too far; we'll delve deeper into deep learning in the next section.

In today's fast-paced world, adaptability is key. Machine Learning algorithms can sift through vast datasets at speeds unattainable by humans, offering invaluable insights and fostering innovative solutions. For businesses, staying competitive often means integrating ML into their operations, from automating customer service with chatbots to enhancing fraud detection in financial transactions.

A noteworthy example of machine learning today is recommendation systems. Think Netflix suggesting the next show you might binge-watch or Amazon recommending products based on your shopping history. These suggestions aren't random; they're the result of complex ML algorithms processing extensive user data to predict preferences and behaviors accurately.

But with great power comes great responsibility. As ML and AI systems become integral to daily life, ethical considerations must be at the forefront of their development. Bias in ML, for instance, can have severe consequences, leading to unfair treatment or discrimination. Thus, creating transparent, fair, and accountable algorithms is more important than ever.

Furthermore, the immense computational power required for effective machine learning poses challenges. While cloud computing and specialized hardware like GPUs and TPUs have mitigated some of these issues, energy consumption and sustainability remain areas needing continual improvement.

Another critical factor is the ongoing need for human oversight. Despite the advancements, machine learning models can still make errors, sometimes with significant ramifications. Ensuring a human-in-the-loop can help catch and correct these mistakes before they escalate.

In the realm of research, universities and tech companies are continually pushing the boundaries of what's possible with ML. From developing algorithms that can write poetry to systems that predict protein folding, the landscape of machine learning is ever-evolving, presenting new opportunities and challenges.

Moreover, the democratization of machine learning tools and platforms means that even small startups and individual developers can now leverage this powerful technology. Open-source frameworks like TensorFlow and PyTorch have lowered the barriers to entry, enabling a wider array of applications and fostering a more inclusive technological revolution.

Critically, Machine Learning is about more than just algorithms and data; it's about creating value. Whether enhancing consumer experiences, streamlining supply chains, or enabling new scientific discoveries, ML holds the promise of unlocking unprecedented possibilities.

Looking ahead, the integration of ML with other emerging technologies, such as Internet of Things (IoT) and blockchain, will likely propel us into a new era of innovation. Imagine smart cities with self-regulating traffic systems based on real-time data or supply chains that autonomously optimize themselves to reduce waste. These are no longer distant dreams but feasible realities being crafted through the convergence of ML and AI.

In summary, Machine Learning is a cornerstone of modern AI, with applications and implications spanning various domains. It's a dynamic field that continually adapts and grows, poised to revolutionize the way we live and work. As we navigate this AI-driven future, staying informed and prepared will be crucial to harnessing the full potential of these technologies.

Deep Learning

Deep learning isn't just a buzzword; it's revolutionizing how we understand and utilize artificial intelligence. At its core, deep learning involves neural networks with many layers—hence the term "deep"—that enable machines to analyze complex data patterns in a way that's almost eerily human. Think of it as giving computers a set of virtual eyes, ears, and even a semblance of intuition. But how did we get here? Well, let's peel back the layers.

The journey begins with neural networks, inspired by the human brain's structure. Early attempts at mimicking neural processes were limited by computational power and data scarcity. Yet, the 21st century brought breakthroughs in both hardware and algorithms. Enter deep learning: a method that leverages vast amounts of data and advanced algorithms to improve machine learning models significantly. Graphical Processing Units (GPUs), originally intended for rendering video game graphics, found a new calling. Their parallel processing capabilities were perfect for the heavy computational lifting required by deep learning.

One of the earliest and most influential neural networks is AlexNet, which won the ImageNet competition in 2012. It wasn't just a victory; it was a revelation. This success showcased the immense potential of convolutional neural networks (CNNs) in image recognition, sparking a wildfire of research and development. Suddenly, industries began to see the practical applications— everything from medical imaging to autonomous vehicles. The ability to recognize, categorize, and interpret images with near-human accuracy opened a floodgate of possibilities.

Deep learning didn't stop at images. Natural language processing (NLP) emerged as another frontier. What if machines could understand and generate human language? This question drove the development of recurrent neural networks (RNNs) and, later, transformers. These models excelled at tasks like language translation, sentiment analysis, and even creative writing. Nowadays, you can chat with AI that crafts emails, stories, and poetry almost indistinguishable from human creation.

For business leaders, deep learning offers unprecedented opportunities. Imagine predictive analytics on steroids. From spotting market trends before they become obvious to optimizing supply chains with laser precision, the potential is staggering. Companies that harness the power of deep learning can outperform their competitors, driving innovation and efficiency to new heights. It's almost like having a crystal ball, but grounded in data and algorithms.

Of course, the road isn't without its bumps. Deep learning models require massive datasets and high-quality data. Garbage in, garbage out. Additionally, the black-box nature of these models—where the decision-making process is often opaque—raises ethical and practical concerns. Transparency in AI decision-making is paramount, especially in critical areas like healthcare and finance. If a neural network makes a mistake, who is accountable? These are not trivial questions, and they need answers as we move forward.

However, deep learning also promises to democratize technology. Open-source frameworks like TensorFlow and PyTorch have lowered the barrier to entry, allowing enthusiasts and small startups to leverage powerful AI tools without needing a supercomputer. This accessibility is fostering a new wave of innovation, with ideas that might otherwise have been relegated to the sidelines now taking center stage.

One area where deep learning is making an indelible mark is healthcare. Imagine training a neural network to read X-rays or MRIs. In some cases, these models have achieved accuracy on par with seasoned radiologists. This isn't about replacing doctors but augmenting their capabilities, allowing for faster and more accurate diagnoses. Personalized medicine is another exciting avenue. Deep learning can help identify patterns in genetic data, paving the way for treatments tailored to an individual's unique genetic makeup.

In the realm of transportation, deep learning is the backbone of autonomous vehicles. Object recognition, decision-making in real-time, and predictive maintenance are all powered by deep learning algorithms. As the technology matures, we're inching closer to a future where self-driving cars are the norm rather than the exception. This shift promises to reduce accidents, optimize traffic flows, and even change the nature of urban planning and development.

However, deep learning isn't a silver bullet. It has its limitations. For one, it can be extremely resource-intensive. Training complex models requires a tremendous amount of computational power and energy. Moreover, deep learning models can and do make mistakes. Understanding the limitations and risks is as crucial as recognizing the potential. Over-reliance without adequate safeguards could lead to disastrous consequences.

There's also the issue of bias. Deep learning models are only as good as the data they're trained on. If the training data is biased, the model's outputs will be too. This has already led to situations where AI systems have exhibited racial and gender biases. As we push the boundaries of what deep learning can achieve, it becomes increasingly important to ensure that ethical considerations keep pace with technological advancements.

The regulatory landscape is beginning to catch up with the rapid advances in deep learning. Policymakers are grappling with questions around data privacy, accountability, and the socio-economic impacts of widespread AI adoption. There's an awareness that regulation must strike a balance—fostering innovation while safeguarding public interests. Businesses must stay abreast of these developments to navigate an ever-evolving landscape.

Looking ahead, the future of deep learning seems boundless. As quantum computing becomes more viable, it promises to supercharge deep learning algorithms, tackling problems previously thought unsolvable. Advanced robotics driven by deep learning could redefine industries, from manufacturing to healthcare, with autonomous systems capable of performing complex tasks with minimal human intervention.

In a world increasingly shaped by artificial intelligence, those who understand and implement deep learning will be at the forefront of innovation. Whether you're a tech enthusiast, a business leader, or just curious about the future, embracing deep learning is akin to harnessing the wind. It offers a competitive edge and opens the door to myriad possibilities. The key lies in leveraging this transformative technology responsibly and ethically, ensuring it serves humanity positively and inclusively.

Chapter 4:
AI in Economics

Imagine harnessing the raw power of artificial intelligence to dissect vast seas of economic data, uncovering trends that were previously invisible to the human eye. AI is revolutionizing how we predict economic shifts, making market analysis not just faster, but astonishingly more accurate. Complex algorithms mine data, spotting patterns and anomalies that can forecast everything from stock prices to global financial crises. In a world where seconds matter, AI offers business leaders and policymakers a precision tool for crafting informed, agile strategies. It's inspiring to think that the very fabric of our economic decision-making is transforming, bringing a future where economic landscapes can be navigated with the finesse of a seasoned sailor reading the stars.

Predictive Analytics

Economics has never been a domain of simple equations and predictable outcomes. It's a complex, often chaotic field influenced by a myriad of variables. To bring some semblance of order to this chaos, analysts have long relied on predictive models. Predictive analytics, powered by AI, represents a quantum leap in this endeavor. By employing advanced algorithms and massive datasets, AI offers unprecedented insights that were, until recently, the stuff of science fiction.

Predictive analytics, at its core, is about making educated guesses about future events based on historical data. In economics, this typically involves forecasting trends, such as consumer spending, market growth, or employment rates. But what sets AI apart from traditional methods is its ability to process enormous amounts of data rapidly and to identify patterns that human analysts might miss.

Consider the manner in which AI can predict financial crises. Traditional models, such as those involving econometric techniques, rely heavily on the analyst's domain knowledge and assumptions. These models can be accurate but are often limited by the analyst's cognitive biases and the simplicity of the models. On the other hand, AI can consider an almost infinite array of variables simultaneously, from macroeconomic indicators to social media sentiment, offering a more holistic and often more accurate prediction.

One practical application of predictive analytics in economics is in the realm of stock market prediction. Machine learning models can be trained on years of historical stock market data, including prices, volumes, and even global news events. These models can then make predictions about future price movements with a higher degree of accuracy than traditional methods. For business leaders and investors, this kind of foresight can provide a significant competitive edge.

Moreover, predictive analytics can play a crucial role in policy-making. Governments can leverage AI to forecast the potential impacts of economic policies before they are implemented. For example, predictive models can simulate the effects of a proposed tax policy on different sectors of the economy, helping policymakers make better-informed decisions. This can lead to more effective and sustainable economic policies that benefit society as a whole.

But let's not get too carried away. While predictive analytics offers numerous advantages, it's not without its challenges. Data quality remains a critical concern. The accuracy of AI predictions is highly dependent on the quality and quantity of the data fed into the algorithms. Incomplete or inaccurate data can lead to flawed predictions, which in turn can lead to poor decision-making. Ensuring data integrity is thus paramount.

Ethical considerations also come into play. Predictive analytics relies heavily on data, much of which is personal in nature. There's a fine line between using data to benefit society and infringing on individual privacy. Businesses and governments must navigate these waters carefully to avoid potential ethical pitfalls.

Then there's the question of interpretability. Complex AI models, such as deep learning networks, are often considered "black boxes" because their decision-making processes are not easily understood even by experts. This lack of transparency can be a significant drawback, especially in a field like economics where understanding the "why" behind a prediction is often as important as the prediction itself. Efforts are underway to develop more interpretable models, but this remains an area in need of further development.

Despite these challenges, the potential benefits of predictive analytics in economics are immense. For instance, consider the realm of supply chain management. Businesses can use AI to predict demand for products more accurately, optimizing inventory levels and reducing waste. In a world where resources are finite, this can lead to more sustainable business practices and contribute to environmental conservation efforts.

It's also worth noting the role of AI in labor markets. Predictive analytics can identify emerging job trends and the skills that will be in demand in the future. This information can help educational institutions tailor their programs to better prepare students for the jobs of tomorrow. It can also assist current workers in upskilling or reskilling, ensuring they remain competitive in an ever-evolving job market.

Even in the volatile realm of cryptocurrencies, predictive analytics is making waves. AI algorithms can analyze the vast amounts of data generated by blockchain transactions, social media, and other sources to predict price fluctuations. Given the high stakes involved, having accurate predictions can make the difference between substantial gains and significant losses.

Another fascinating application is in agriculture. Farmers can use predictive analytics to forecast crop yields based on various factors such as weather conditions, soil quality, and historical data. This can lead to more efficient farming practices and better food security. In this way, AI has the potential to revolutionize even the most traditional sectors of the economy.

It's not all about profit, either. Non-profit organizations and social enterprises can leverage predictive analytics to optimize their impact. For example, AI can help predict areas of economic hardship and direct resources where they are needed most. This can lead to more effective interventions and ultimately improve the quality of life for many people.

In the world of real estate, predictive analytics can offer valuable insights into property price trends. By analyzing historical sales data, demographic shifts, and economic indicators, AI can help investors make more informed decisions. This can lead to more stable and profitable investments, contributing to overall economic stability.

The potential applications of predictive analytics in economics are virtually limitless. However, it's essential to approach this technology with a balanced perspective. While AI offers powerful tools for prediction and analysis, it's not infallible. Human oversight remains crucial to ensure that predictions are not only accurate but also ethical and aligned with broader societal goals.

In conclusion, as we stand on the cusp of an AI-driven future, the role of predictive analytics in economics cannot be overstated. It offers the promise of more accurate forecasts, better decision-making, and ultimately, a more prosperous and sustainable world. However, realizing this promise will require ongoing efforts to address the challenges of data quality, interpretability, and ethics. By doing so, we can harness the power of AI to create a brighter economic future for all.

Market Trends

The integration of artificial intelligence (AI) in economics isn't just a theoretical exercise; it's a dynamic force that's already reshaping markets around the globe. At the heart of this transformation are the emerging market trends that reflect both the potential and challenges of AI technologies. From stock market forecasting to the emergence of new business models, AI's footprint in economics is undeniably expanding.

One of the most profound trends is the rise of **algorithmic trading**. Traditionally, stock trading was a domain dominated by human intuition and analysis. Today, AI-driven algorithms can analyze vast amounts of data at lightning speed, making split-second decisions that capitalize on market inefficiencies. These systems aren't confined to Wall Street alone; they're finding applications across global markets. Surprisingly, algorithms can process a combination of historical data, real-time news, and even social media sentiment to predict market movements. *It's like having a crystal ball with a processor chip.*

Yet, the prominence of algorithmic trading brings its own set of challenges. Flash crashes, like the one experienced in 2010, have sparked debates on the need for regulations and safety nets in an AI-dominated trading environment. As these trends mature, the conversation around robust governance and ethical use of AI becomes vital.

Another area witnessing a significant uptick due to AI is **financial forecasting**. Businesses and investors alike are leveraging AI-powered predictive analytics to drive decision-making. Whether it's projecting quarterly revenues, forecasting market demand, or even predicting bankruptcy risks, AI has revolutionized economic forecasting. Traditional models relied heavily on assumptions and human judgment, often falling short in volatile markets. AI, on the other hand, adapts and learns from each data point, constantly refining its predictions.

Let's talk about consumer behavior analytics. Understanding consumer behavior has always been a priority for businesses, but the tools available today are lightyears ahead of traditional methods. AI systems analyze buying patterns, social media interactions, and even browsing histories to generate nuanced consumer profiles. Retail giants and startups alike use these insights to tailor their marketing strategies, optimize product placements, and even manage inventory efficiently. Personalized marketing campaigns driven by AI can now target individuals based on their unique preferences, rather than broad demographic categorizations.

Moreover, the gig economy is feeling AI's touch in unexpected ways. Platforms like Uber, Airbnb, and TaskRabbit are using AI to match supply with demand more effectively. Predictive algorithms help these platforms optimize pricing, project future demand, and allocate resources efficiently. This can mean dynamic surge pricing during peak hours or intelligently recommending the best tasks or rentals to users.

AI's impact on sustainable economic practices is another exciting trend. For instance, in agriculture, AI technology helps in optimizing yields by predicting the best times for planting and harvesting. In manufacturing, AI is used for predictive maintenance – analyzing equipment data to predict failures before they happen. Both these applications lead to more efficient use of resources and cost savings, contributing to a sustainable economy.

Digital currencies and blockchain technology have also been swept up in the AI tide. Cryptocurrencies like Bitcoin and Ethereum are highly volatile, making them perfect candidates for AI-driven market analysis and trading. AI algorithms can process transaction data, detect fraudulent activities, and even predict future trends based on market sentiment. This cross-pollination of AI and blockchain technology is producing a fertile ground for financial innovation, with new startups emerging that combine both technologies to offer groundbreaking solutions.

The labor market isn't left out either. AI is being used to analyze job market trends and predict the skills that will be in high demand. Platforms like LinkedIn use AI to match job seekers with potential employers more effectively. They analyze a plethora of data points, from job seeker profiles to market demand, helping businesses and candidates find better fits. This trend suggests that while AI may displace certain job categories, it will also create opportunities by spotlighting emerging fields and required skillsets.

While AI's contributions to the economy are generally beneficial, it's not all smooth sailing. Disparities in AI adoption and deployment can create economic divides. Companies and countries that can afford to implement cutting-edge AI systems gain a competitive advantage, potentially widening the gap between the technologically advanced and the laggards. This, in turn, raises important questions about equitable access to AI technologies and the need for inclusive growth strategies.

Moreover, the increasing prevalence of AI systems necessitates new frameworks for data governance and ethical considerations. As AI systems access more personal and corporate data, the demand for robust cybersecurity measures becomes more pressing. Data breaches and misuse of AI-driven insights can have far-reaching implications for businesses and economies alike. Crafting policies and standards to govern this data usage is, therefore, an emerging trend that cannot be ignored.

We also see a growing trend of interdisciplinary collaboration. Economists, data scientists, and technologists are coming together to create more sophisticated AI models that can tackle complex economic questions. This collaboration is crucial because it ensures that the economic models built using AI are grounded in sound economic principles while leveraging cutting-edge technology.

Finally, venture capital interest in AI-driven economic solutions is skyrocketing. Investors are increasingly looking for startups that use AI to address unique challenges within the economic landscape. Whether it's financial technology (FinTech), retail analytics, or supply chain optimizations, the diversity of applications is vast, and the potential for impact is huge. This influx of capital not only propels technological advancements but also spurs job creation and economic growth.

In conclusion, the market trends emerging from AI's integration into economics are as vast as they are varied. From algorithmic trading to consumer behavior analytics, and financial forecasting to sustainable practices, AI is proving to be a transformative force. To navigate this AI-driven future, businesses, policymakers, and individuals will need to stay informed and adapt strategies accordingly. The future of economics in an AI world promises both unprecedented opportunities and challenges, making now the time to engage, innovate, and lead.

Chapter 5:
AI in Healthcare

A I in healthcare is nothing short of transformative, unfolding a future where diagnosis is precise and treatment is personalized. Imagine a world where AI algorithms analyze vast amounts of medical data in seconds, providing doctors with valuable insights and even predicting the onset of diseases before symptoms appear. Training these algorithms on millions of patient records, AI can identify patterns and anomalies that human eyes might miss, making early diagnosis and intervention more common. Personalized medicine takes this a step further by tailoring treatments to an individual's unique genetic makeup, lifestyle, and environment, maximizing efficacy and minimizing side effects. AI doesn't just enhance operational efficiencies and improve patient outcomes; it revolutionizes healthcare as we know it. It's a ground-breaking journey promising not only to extend human life but also to significantly improve its quality.

Diagnostic Tools

In the realm of healthcare, one of the most transformative applications of AI is in diagnostic tools. These remarkable systems are revolutionizing the way medical professionals identify and treat diseases, propelling us into a new era of precision medicine. You might be wondering how AI is doing this. It's essentially a blend of computational algorithms and vast datasets, combined to read medical images, analyze patient data, and even make predictive assessments.

One compelling example of AI's prowess in diagnostics is its ability to detect anomalies in medical imaging. MRI scans, X-rays, and CT images are traditionally scrutinized by radiologists. It's a meticulous process requiring significant expertise and time. Enter AI—a well-trained machine learning model can now examine thousands of these images in a fraction of the time and often with higher accuracy. Studies have demonstrated that AI systems can detect conditions such as tumors, fractures, and even subtle vascular irregularities that might evade the human eye.

The implications of these advancements are stunning. Faster and more accurate diagnostics lead to quicker interventions. For patients, this means better outcomes and, in some cases, lives saved. For healthcare systems, it translates to reduced costs and increased efficiency. Imagine being able to cut down the time to diagnose a critical condition from days to minutes; that's not just transformative—it's revolutionary.

Take, for instance, the detection of diabetic retinopathy, a common eye disease among diabetes patients. Traditionally, detecting this condition requires a comprehensive eye examination, typically conducted by highly trained specialists. AI systems, however, can quickly screen for signs of diabetic retinopathy from fundus photographs with a remarkable level of accuracy. This technology is expanding access to essential diagnostic services, especially in under-resourced regions where specialists are scarce.

AI is also making strides in genomics, opening new frontiers in our understanding of diseases at the molecular level. Bioinformatics algorithms are now capable of analyzing massive datasets of genetic information to identify biomarkers for various health conditions. These biomarkers can not only alert doctors to the presence of diseases but also offer insights into how a patient might respond to certain treatments. It's personalized medicine at its finest.

The use of AI in diagnostics isn't just limited to imaging and genomics. Natural language processing (NLP) is another burgeoning field. Medical histories, clinical notes, lab reports—these are treasure troves of information. Yet, extracting useful insights from them can be a Herculean task for humans. NLP algorithms can sift through heaps of unstructured data, summarizing relevant information and even identifying correlations that might not be immediately apparent. This augments the doctor's ability to make informed decisions swiftly.

Moreover, AI-powered diagnostic tools are becoming increasingly integrated into wearables and mobile apps, making healthcare more accessible than ever before. Think about wearables that continuously monitor your heart rate, sleep patterns, and other vital signs. These devices can now utilize AI to detect irregularities that might suggest underlying health conditions. Alerts can be sent directly to your medical provider, enabling preemptive care.

Consider the digital stethoscope, a device that can record heart sounds and analyze them using AI. These systems can identify murmurs and other abnormalities, providing a preliminary diagnosis before a physician even reviews the data. Similarly, smartphone apps equipped with AI capabilities can now analyze skin lesions to differentiate between benign and malignant conditions.

However, while the potential is immense, it's crucial to address the ethical, regulatory, and practical challenges that come with integrating AI into diagnostic processes. Accountability remains a pressing issue— who is responsible if an AI system makes an incorrect diagnosis? Moreover, ensuring data privacy and security is vital. These tools rely on copious amounts of personal medical data, which must be handled with the utmost care.

Regulatory bodies across the globe are grappling with how best to oversee the proliferation of AI in diagnostics. Rigorous validation and continuous monitoring of AI systems are essential to maintain their reliability and trustworthiness. That said, partnerships between technology developers, healthcare providers, and regulatory agencies are paramount to navigate these challenges effectively. This collaborative approach will ensure that AI systems enhance, rather than hinder, patient care.

But let's not lose sight of the human element. AI is a tool—an extraordinarily powerful one—but it cannot replace the empathy and nuanced understanding of human physicians. Instead, think of AI as an assistive partner, augmenting the capabilities of healthcare professionals. By automating routine tasks and providing deeper insights, AI allows doctors to focus more on patient care, improving the overall experience.

In conclusion, the advent of AI-powered diagnostic tools is paving the way for a more efficient, accurate, and personalized healthcare system. From enhancing image analysis to driving forward genetic research, the applications are as diverse as they are impactful. And while challenges remain, the promise of improved patient outcomes and a more responsive healthcare system makes the integration of AI in diagnostics an exciting frontier worth pursuing.

As we continue to refine these technologies and develop new applications, we stand on the brink of a healthcare revolution. One where timely, accurate diagnoses are the norm rather than the exception. One where AI aids in unlocking the full potential of medical science, ensuring that quality care becomes universally accessible.

Indeed, the future of diagnostics in healthcare looks bright, bolstered by the relentless advancements in artificial intelligence. As we forge ahead, the symbiosis between human ingenuity and machine intelligence will undoubtedly redefine the contours of medical diagnostics, leading us into an era of unparalleled precision and care.

Personalized Medicine

Imagine a world where your medical treatment isn't based on generalized statistics but tailored specifically to you. That's the crux of personalized medicine. Thanks to the rapid advancements in artificial intelligence, this dream is turning into reality. Personalized medicine leverages AI algorithms to analyze vast datasets, including genetic information, health records, and even lifestyle factors, to devise individualized treatment plans.

At the heart of this transformation is the ability of AI to process enormous amounts of data quickly and accurately. By analyzing genetic sequences, AI can identify specific mutations responsible for diseases. This analysis enables healthcare providers to prescribe treatments that target these unique genetic abnormalities effectively. For instance, AI can help determine the optimal chemotherapy drug for a cancer patient based on the genetic profile of their tumor, increasing the chances of a successful outcome.

Moreover, AI-powered tools are facilitating early diagnosis and prevention strategies. These tools can track subtle changes in a person's health data, identifying potential issues before they become significant problems. It's a proactive approach that shifts the focus from treating diseases to preventing them. Imagine receiving alerts on your smartphone that prompt you to see a doctor because your routine biometrics indicate a potential health risk. That's personalized medicine in action.

The potential benefits go beyond better treatment outcomes. Personalized medicine can also reduce side effects and improve patients' quality of life. Traditional treatments often have a one-size-fits-all approach, leading to varying responses among patients. With AI, we can minimize these variations by selecting the most effective medication with the least side effects for each individual. This precision not only enhances the therapeutic experience but also optimizes healthcare resources, reducing overall costs.

Think about how transformative this is for chronic conditions like diabetes or hypertension. AI algorithms can continuously monitor patient data through wearable devices, adjusting treatment plans in real-time based on the collected information. This dynamic feedback loop ensures that the patient is always receiving the most effective care, adapting to changes as they occur. No more waiting for the next doctor's appointment to adjust your medication; AI handles it instantly.

Another exciting aspect is the role of AI in genomics. The human genome is incredibly complex, with over three billion base pairs of DNA. Sequencing and analyzing this data manually would be an insurmountable task. However, AI can sift through these sequences to identify patterns and correlations that might be missed by human eyes. This capability opens the door for breakthroughs in understanding genetic predispositions to various diseases and developing corresponding treatments.

As we delve into the realm of rare diseases, the advantages of personalized medicine become even more pronounced. Rare diseases are often overlooked due to their low prevalence, leading to delays in diagnosis and treatment. AI can combat this by recognizing rare disease patterns in patient data, ensuring timely and accurate diagnoses. This early intervention can be lifesaving for many who suffer from conditions that have traditionally received less attention.

The integration of AI in personalized medicine isn't limited to physical health. Mental health is another domain where personalized approaches are making significant strides. AI algorithms can analyze data from various sources such as social media, wearable devices, and even voice recognition software to identify signs of mental health issues. By understanding an individual's unique triggers and stressors,

AI can recommend personalized coping strategies or interventions, making mental health care more effective and accessible.

However, the adoption of personalized medicine isn't without challenges. Data privacy is a significant concern. Patients' genetic information is incredibly sensitive, and safeguarding this data is paramount. AI solutions must incorporate robust security measures to protect patient information from breaches and misuse. Moreover, there's the ethical aspect of how this data is used. Clear guidelines and regulations are necessary to ensure that AI in personalized medicine serves the best interests of patients while respecting their privacy.

Interoperability of healthcare systems is another hurdle. For AI to provide truly personalized care, it needs access to a comprehensive set of patient data. This requirement necessitates seamless integration across various healthcare providers and electronic health record systems. The good news is that many stakeholders recognize this need and are working towards standardized protocols to facilitate data sharing.

Education and training of healthcare professionals is yet another critical piece of the puzzle. Doctors, nurses, and other healthcare workers need to understand how to interpret AI-generated insights and integrate them into their clinical practice. This shift requires a blend of technological proficiency and traditional medical knowledge, emphasizing the importance of continuous learning and adaptability in the healthcare sector.

Financial implications also play a role in the widespread adoption of personalized medicine. Developing and implementing AI technologies require substantial investment. While the long-term benefits may justify the costs, initial funding can be a barrier, especially in resource-constrained settings. Policymakers and healthcare leaders must strategize on funding models that ensure equitable access to personalized care for all patients, regardless of their financial background.

In conclusion, personalized medicine represents a seismic shift in how we approach health and wellness. With AI at the helm, we are moving towards a future where treatments are as unique as the individuals receiving them. This transformation promises not just better health outcomes but also a more efficient and equitable healthcare system. As tech enthusiasts, business leaders, and forward-thinking individuals, it's essential to embrace and advocate for this evolution, ensuring that the benefits of AI-driven personalized medicine reach everyone. The journey is just beginning, and the potential for innovation is boundless.

Chapter 6:
AI and Personal Privacy

Moving from the lifesaving potential of AI in healthcare to its intricate dance with personal privacy, we find ourselves in a realm where the stakes are equally high but shaded with ambiguity. In a world where data streams faster than oil gushes from wells, our private lives are increasingly sketched in binary code. AI systems, eager to optimize user experiences, can also tiptoe into areas of our lives we cherish as private. The issue isn't just about data breaches; it's about how much of our personal data should be harvested in the first place. Balancing AI's capabilities with personal privacy isn't just a technical challenge—it's an ethical maze, requiring businesses and lawmakers to join forces in mapping out new boundaries. As we harness AI's power, we must not lose sight of the fundamental right to privacy. This chapter aims to provide a comprehensive overview of this critical intersect, serving as a guiding star for navigating the murky waters of data security and ethical considerations.

Data Security

In an era where artificial intelligence (AI) seeps into almost every facet of our lives, data security stands as a pillar of paramount importance. Whether it's financial institutions leveraging AI for fraud detection or healthcare providers using machine learning algorithms to enhance diagnostic accuracy, the common thread is the massive amounts of data being handled. And where there's data, there's a pressing need for secure handling, storage, and transfer to prevent unauthorized access and breaches.

At its core, data security revolves around protecting data from corruption, theft, or unauthorized access. In the context of AI, this takes on new layers of complexity because AI systems require vast amounts of data to function effectively. The richer and larger the dataset, the more potent the AI system can be. However, this treasure trove of data is also a magnet for cybercriminals.

Let's dive into some key strategies for ensuring data security in an AI-driven world. First, encryption is a critical shield. By converting data into cryptographic code, encryption makes it unreadable to anyone who doesn't have the proper decryption key. It's like having a secret language that only you and your intended recipient can understand. Encryption should be applied both to data at rest (stored data) and data in transit (data being transferred across networks).

Data anonymization is another technique that's gaining traction. Here, personally identifiable information (PII) is either removed or encrypted. This way, even if data falls into the wrong hands, it isn't attached to identifiable individuals. Companies can still use this 'de-identified' data for valuable insights without compromising personal privacy.

But encryption and anonymization are just parts of the puzzle. Access controls play a crucial role in safeguarding data. By implementing stringent protocols about who can access the data and what they can do with it, organizations can significantly reduce the risk of insider threats. Multi-factor authentication (MFA) and role-based access controls (RBAC) are common practices in this realm.

Beyond these technical measures, the culture of the organization matters. Instilling a strong sense of data privacy and security within the workforce can substantially mitigate risks. Regular training sessions that highlight the importance of data security and provide best practices go a long way. Humans, after all, are often the weakest link in cybersecurity defenses.

Let's talk about secure data storage. The cloud has revolutionized the way we store data, but it has also introduced new vulnerabilities. Cloud providers offer robust security frameworks, but it's a shared responsibility. Organizations should ensure their chosen cloud service providers adhere to strict compliance standards and encryption practices. Additionally, adopting a hybrid approach, combining on-premises and cloud storage, can offer an extra layer of protection.

Blockchain technology is another fascinating development in the landscape of data security. Initially designed for cryptocurrencies, its decentralized nature ensures that data is stored across a network of computers. This decentralization makes it nearly impossible for any one entity to manipulate the data without consensus from the rest of the network. Incorporating blockchain into AI systems can add an extra layer of transparency and security.

Audits and compliance checks should also be part of the playbook. Regular internal and external audits can identify vulnerabilities that need fixing. Being compliant with regulations like the General Data Protection Regulation (GDPR) and the California Consumer Privacy Act (CCPA) isn't just a legal obligation—it's a commitment to upholding the highest data security standards.

Let's not forget about the role of AI in enhancing data security itself. Yes, AI can be both a protector and an adversary in the data security narrative. Machine learning algorithms can analyze patterns and detect anomalies in real-time, offering a proactive approach to identifying potential security breaches. These systems can adapt and learn from each incident, becoming more robust over time.

Nonetheless, the double-edged sword aspect of AI must be acknowledged. The very AI technologies that promise enhanced security can also be co-opted by malicious entities to orchestrate more sophisticated attacks. This cat-and-mouse game between cyber defenders and attackers is ever-present, driving the continuous evolution of security measures.

Data security is not merely a technical challenge but also a pressing ethical concern. Mishandling data or failing to secure it not only leads to potential financial losses and legal ramifications but also breaks the trust of users. Trust, once lost, is incredibly hard to regain. Therefore, data security measures should be as robust as the AI systems they aim to protect.

In closing, navigating the intersection of AI and data security is like walking a tightrope. On one side, you have the incredible potential of AI to revolutionize industries and improve lives. On the other, you have the ever-present threat of data breaches and cyber-attacks that can derail these advancements. The key lies in a balanced approach that combines cutting-edge technology with rigorous security protocols and ethical considerations.

So, as we forge ahead into this uncharted territory, let's not lose sight of the foundational importance of data security. It's not just a technical necessity; it's the bedrock upon which the future of AI will be built. And that future, in many ways, hinges on our ability to keep data safe and secure.

Ethical Considerations

As AI technology continues to evolve, the ethical considerations surrounding its use become increasingly complex. Balancing the benefits of AI with the need to protect personal privacy is no small feat. It requires a multifaceted approach that takes into account various ethical principles, social norms, and legal frameworks. So, what does this ethical quagmire entail, especially when it comes to personal privacy?

One of the most pressing ethical dilemmas involves the collection and use of personal data. AI systems rely heavily on large datasets to function effectively. Whether it's for training machine learning models or developing deep learning algorithms, data is the fuel. However, this raises questions about the ownership of data. Who owns your data once it's collected? The individual, the company collecting it, or the AI system itself?

Transparency is another critical ethical consideration. In an ideal world, AI systems would operate in a fully transparent manner, allowing individuals to understand how their data is being used and for what purposes. Unfortunately, many AI models are seen as "black boxes," making it difficult for even their creators to fully explain their internal workings. This lack of transparency can erode public trust and lead to misuse or abuse of personal data.

Moreover, we must also consider the consent of individuals whose data is being collected and analyzed. Informed consent should be an essential component of any data collection endeavor. Yet, how often do people actually read and understand the terms and conditions presented to them? The practice of burying crucial information in lengthy documents must be re-evaluated. AI-driven privacy policies should be more accessible and understandable to the everyday user.

But let's not just look at the negatives. Ethical considerations also involve leveraging AI for social good. Properly designed AI systems can significantly enhance personal privacy by discovering and mitigating breaches more swiftly than any human could. Imagine an AI watchdog that alerts you whenever your data is being misused or sold without your consent. Such systems could empower individuals to reclaim control over their personal information.

Then, there's the issue of bias. AI systems are only as good as the data they are trained on. If the data contains biases, the AI will inevitably perpetuate these biases, leading to unfair and discriminatory outcomes. This becomes particularly troublesome when it comes to personal privacy and data security. Biased algorithms can disproportionately impact certain groups, leading to a greater invasion of their privacy.

The ethical implications also reach into the realm of data security. How do we ensure that AI systems are secure and that they protect the data they are entrusted with? Using state-of-the-art security measures is crucial, but it should not end there. Regular audits and rigorous testing must be conducted to identify vulnerabilities and potential breaches. Ethical AI development mandates an ongoing commitment to fortifying these systems against malicious activities.

Accountability is another cornerstone. When AI systems make decisions that affect individuals, who is held accountable? Is it the developer, the company that deploys the AI, or the AI system itself? Establishing clear lines of responsibility is essential to address the ethical considerations adequately. Companies that develop and deploy AI systems must adopt a framework that ensures accountability at all levels, ensuring any misuse or ethical violations are promptly addressed.

Does the end justify the means? While AI can bring about phenomenal advancements and efficiencies, we must question whether these benefits are worth the cost to personal privacy. This ethical consideration often boils down to the trade-offs we're willing to make. Would you trade your personal data for more convenience? Are you comfortable with a machine knowing more about you than your closest friends and family?

Yet, it's not all doom and gloom. Ethical considerations also drive innovation in AI. As we grapple with these issues, new solutions emerge. For example, evolving norms around data minimization aim to collect only the data that is truly necessary. Federated learning techniques are gaining traction, allowing algorithms to be trained on data without that data ever leaving its original location. Such innovations showcase the possibility of ethical AI that respects personal privacy.

Finally, education is vital. Stakeholders—ranging from developers and policymakers to end-users—must be well-versed in the ethical implications of AI. Ethical considerations shouldn't just be an afterthought; they should be an integral part of the AI development lifecycle. Educating all parties about the ethical challenges and potential solutions can help create a more informed and responsible approach to AI.

In the grand scheme of things, ethical considerations surrounding AI and personal privacy are intricate and multifaceted. They demand our continuous attention and thoughtful action. Striking the right balance between leveraging AI's capabilities and safeguarding our personal privacy will be challenging but far from impossible. The path forward involves transparent practices, robust security measures, informed consent, reducing biases, and a shared sense of accountability.

Chapter 7:
AI in Business

In the heart of today's business revolution, AI stands as a formidable catalyst for transformation, promising operational efficiency that stretches beyond traditional benchmarks while personalizing customer experiences like never before. Imagine systems that can predict market demands, optimize supply chains, and automate repetitive tasks, leaving human creativity to tackle more meaningful challenges. This isn't some far-off future—it's already here, proving indispensable from retail giants to small startups. Adopting AI means shifting from reactive to proactive strategies, leveraging data not just to enhance but to truly understand consumer behavior. For business leaders, the journey involves not just integrating cutting-edge technologies but reimagining their roles and capacities. At its core, AI in business isn't just about smarter algorithms; it's about fostering a culture ready to embrace continuous innovation and radical change.

Operational Efficiency

Operational efficiency is more than just a buzzword; it's the lifeblood of successful businesses. At its core, operational efficiency is about making the most of your company's resources—time, money, and manpower—to provide better products or services at lower costs. AI brings a transformative edge to this endeavor, enabling organizations to achieve unprecedented levels of efficiency. Whether it's streamlining supply chains, optimizing routes for delivery, or automating repetitive tasks, AI has the potential to revolutionize how businesses operate.

One of the most compelling features of AI in operational efficiency is its ability to process vast amounts of data quickly. Businesses are awash with data, but data alone isn't enough; it's the actionable insights derived from that data that drive efficiency. By employing machine learning algorithms, for example, companies can detect patterns and predict outcomes. This allows for smarter decision-making processes that reduce waste and increase productivity.

Take inventory management, a crucial aspect of operational efficiency. Traditional inventory management systems can be cumbersome, bogged down by manual data entry and prone to errors. AI, however, can automate these tasks, identifying trends and predicting future needs with startling accuracy. You can imagine a warehouse where robotic systems are not only responsible for picking and packing but are also continually optimized by AI to enhance their performance. The result? Reduced human error, lower operational costs, and faster delivery times.

AI's impact on operational efficiency isn't confined to back-end processes. Front-end tasks, such as customer service, have also seen remarkable improvements. Intelligent chatbots and virtual assistants can handle a wide range of customer queries without the need for human intervention. Moreover, they can operate 24/7, providing customers with instant solutions and freeing up human agents to tackle more complex issues. Consider how this shift not only boosts customer satisfaction but also helps in reallocating human resources to more strategic roles.

Furthermore, predictive maintenance offers perhaps one of the most tangible benefits of AI in operational efficiency. In industries reliant on machinery—whether it's manufacturing, transportation, or utilities—the ability to predict when a machine might fail can save millions. Instead of waiting for a breakdown, AI algorithms can analyze sensor data to forecast equipment failures, allowing for timely maintenance. This not only prolongs the lifespan of the machinery but also minimizes downtime, thus maintaining a smooth operational flow.

For supply chains, the game-changing aspect of AI comes through optimization. Supply chains are usually complicated networks with multiple touchpoints—which makes them ripe for inefficiencies. AI can sift through mountains of data to identify where bottlenecks occur and suggest improvements. Sophisticated algorithms can predict demand, streamline logistics, and even negotiate better supplier contracts. By providing real-time insights, AI helps companies respond more quickly to market changes, enhancing agility and resilience.

The power of AI in operational efficiency extends to workforce management as well. Scheduling shifts, managing absences, and optimizing the allocation of tasks can be a logistical nightmare. AI systems can analyze historical data to predict when you might need more staff and where. This ensures that the right number of people are always in the right place, performing the right tasks. As a result, businesses can significantly improve productivity while also boosting employee satisfaction through better-managed workloads.

Energy management is another area where AI shines. Businesses are under increasing pressure to reduce their carbon footprints, and efficient energy use is a critical part of this. AI-driven systems can monitor energy consumption in real-time, offering insights into how to use energy more efficiently. They can automatically adjust heating, cooling, and lighting based on occupancy and usage patterns, significantly reducing energy costs and environmental impact.

But what about industries like healthcare and finance, where operational efficiency can have a direct impact on lives and livelihoods? In healthcare, AI can streamline administrative workflows, such as scheduling and patient record management, freeing up more time for patient care. Meanwhile, in finance, algorithms can automate fraud detection, risk management, and even trading activities. Both sectors benefit from a reduction in manual, time-consuming tasks, allowing professionals to focus on higher-value activities.

And it's not just large enterprises that stand to gain. Small and medium-sized businesses can also leverage AI to enhance operational efficiency. Cloud-based AI services make sophisticated tools accessible without the need for heavy upfront investment. From social media marketing and customer relationship management to inventory and HR management, these tools can help smaller businesses compete on a level playing field with larger companies.

Furthermore, employee training and development benefit tremendously from AI-driven insights. Skills gaps can be identified, and personalized training programs can be developed to upskill the workforce efficiently. By continuously monitoring performance metrics, AI ensures that training is effective and that employees are always equipped with the skills needed to meet evolving business demands. This turns talent management into a proactive rather than reactive process, making the organization more agile and adaptable.

Some might argue that the implementation of AI systems is costly and complex. While there are initial investments involved, the long-term gains far outweigh these costs. Not only does AI streamline operations, but it also provides a competitive edge that is becoming increasingly essential in today's fast-paced market. The key is to approach AI implementation thoughtfully, focusing on areas where it can deliver the most value and scaling up from there.

The human element should not be overlooked in the rush toward greater automation. The most effective AI systems operate in tandem with human intuition and expertise, forming a symbiotic relationship that augments rather than replaces human roles. Employees should be seen as valuable assets in this transformation, equipped and empowered to work alongside AI for greater efficiency.

The path to operational efficiency through AI is not a straightforward one, but its possibilities are limitless. Embracing AI involves a cultural shift, a willingness to innovate, and an openness to adapt. By doing so, businesses set themselves on a course not only for enhanced operational efficiency but also for sustainable growth and long-term success. As AI continues to evolve, so too will its capabilities, offering ever more sophisticated ways to refine and improve the efficiency of business operations.

Customer Experience

Customer experience is quickly becoming one of the most transformative areas in which AI is making its mark on the business landscape. In a world where consumer expectations are continually rising, companies are leveraging AI to meet these demands in more efficient, personalized, and impactful ways. It's not just about selling products anymore; it's about creating an experience that resonates with the customer on an emotional level.

At the heart of AI-driven customer experience lies the concept of personalization. Today's consumers expect companies to understand their needs and preferences, sometimes even before they do. AI enables businesses to analyze vast amounts of data, glean insights, and deliver personalized experiences that can lead to increased customer loyalty and satisfaction. Imagine a retail website that can recommend products based on your browsing history, purchase patterns, and even current mood—all thanks to sophisticated algorithms working behind the scenes.

But it's not just online platforms that are benefiting from AI. Brick-and-mortar stores are also finding innovative ways to integrate AI into their customer service strategies. Retailers can use AI-powered tools to manage inventory more efficiently, ensuring that popular products are always in stock and ready for customers. AI can also help create seamless checkout experiences, reducing wait times and making the shopping experience as pleasant as possible.

One of the standout applications of AI in customer experience is through chatbots and virtual assistants. These intelligent systems can handle a range of customer queries and issues, often resolving them in real-time. Not only does this reduce the burden on human customer service agents, but it also ensures that customers receive quick and accurate responses, enhancing their overall experience. The best part? These systems are available 24/7, meaning that customer support is always within reach.

Consider the impact of AI on customer feedback mechanisms. Traditional methods of gathering customer feedback can be time-consuming and often don't provide a real-time snapshot of customer sentiment. AI can analyze customer reviews, social media comments, and other forms of feedback instantly, identifying patterns and trends that might have gone unnoticed. Companies can then act on this information more swiftly, making adjustments that improve customer satisfaction and loyalty.

In the realm of marketing, AI is revolutionizing how companies reach and engage with their customers. AI-powered analytics can determine the best times to send marketing emails, the type of content that will resonate most with a specific audience, and even the ideal frequency of interactions. By making data-driven decisions, businesses can optimize their marketing efforts to reach the right people at the right time with the right message. This not only maximizes ROI but also ensures that customers receive content that is relevant and engaging.

Another aspect of AI-driven customer experience is predictive maintenance in industries like automotive and electronics. By analyzing data from past service records, AI can predict when a particular component is likely to fail and send notifications to customers, suggesting preemptive maintenance. This proactive approach not only extends the lifespan of products but also builds trust between the customer and the brand, as it shows a commitment to their long-term satisfaction.

When we look at the potential for AI in customer experience, it's clear that we're just scratching the surface. The future holds even more possibilities, from augmented reality (AR) shopping experiences to AI-driven customer journey mapping that can predict and influence buying behaviors. Imagine walking into a store where AR glasses allow you to see product information, reviews, and even virtual try-ons, all tailored to your preferences and past behaviors.

Let's explore a compelling example: Imagine a hotel chain using AI to enhance guest experiences. Upon booking, the AI system begins to collect data on the guest's preferences, past stays, and even their social media activity. By the time the guest checks in, the room is customized to their liking—preferred room temperature, favorite snacks, and even personalized entertainment options are ready to go. During the stay, the AI concierge can recommend activities, dining options, and even suggest the best time to visit popular local attractions based on current traffic and crowd data.

In essence, AI is transforming customer experience from a reactive to a proactive process. Businesses are no longer waiting for customers to state their needs; they are anticipating them and delivering solutions in real time. This shift is not just enhancing customer satisfaction but also creating new standards for what consumers expect from businesses. Companies that fail to adapt risk being left behind as the bar for customer experience continues to rise.

The ethical considerations of using AI in customer experience also deserve attention. While personalization can greatly enhance the customer experience, it can also lead to concerns about privacy and data security. Companies must balance the benefits of AI with the responsibility to protect customer data and be transparent about how it is used. Trust is a critical component of customer experience, and mishandling data can erode that trust, leading to long-term damage to the brand.

AI's role in customer experience is not just an enhancement but a necessity in today's competitive business environment. As AI technology continues to evolve, so too will the methods by which companies interact with their customers. From predictive analytics to real-time personalization, the potential for AI to redefine customer experience is immense and the journey has only just begun.

In conclusion, the integration of AI into customer experience strategies is revolutionizing how businesses operate and interact with their consumers. Companies that harness the power of AI to anticipate and meet customer needs will not only thrive but set new benchmarks in customer satisfaction and loyalty. As we move forward, the fusion of human empathy and AI efficiency could very well become the gold standard in delivering exceptional customer experiences.

Chapter 8:
AI and the Workforce

Artificial Intelligence is shaking up the workforce in ways we've only begun to imagine. As algorithms become smarter and machines more capable, debates around job displacement and newly created roles heat up. Picture a world where repetitive, mundane tasks are handled by machines, freeing human ingenuity to tackle more complex, creative challenges. But don't think for a second it's all smooth sailing. There are real concerns about job loss and the need for radical shifts in skillsets. Sure, some jobs will vanish, but countless new opportunities will also arise—roles that don't even exist yet. It's a thrilling, nerve-wracking balancing act, walking the tightrope between innovation and disruption. The future of work is less about pushing papers and more about leveraging human insight in ways we never thought possible. It's an evolving narrative—one where advanced technology and human potential coalesce to redefine what it means to work, thrive, and grow in the digital age.

Job Displacement

As artificial intelligence continues to evolve, its impact on the workforce is becoming increasingly significant. One of the most pressing concerns is job displacement. It's crucial to understand how AI is changing the job landscape.

Job displacement due to AI isn't a new discussion. From steam engines to the internet, technological advancements have always spurred waves of job loss and creation. AI, however, is a revolution of a different magnitude because of its ability to perform tasks that were once thought to be the exclusive domain of humans.

AI can analyze vast amounts of data in seconds, automate routine tasks, and even perform complex activities that require a degree of learning and adaptation. Jobs that involve repetitive tasks, data entry, and even certain aspects of customer service are high on the list of those at risk. Imagine factory floors running smoothly with almost zero human intervention or customer service queries being resolved efficiently by chatbots. This kind of automation is already happening and will become more sophisticated with time.

Experts often cite the "Fourth Industrial Revolution" in conversations about AI and job displacement. Think of it this way: robotics and AI are to the twenty-first century what the steam engine was to the nineteenth. The ramifications are broad and varied, affecting blue-collar and white-collar jobs alike. From truck drivers to financial analysts, no sector is entirely immune.

While this might sound daunting, it's essential to recognize that AI also offers opportunities. For instance, even as traditional roles diminish, new types of jobs are emerging. Careers such as data scientists, AI specialists, and technology ethicists are in high demand. These roles largely didn't exist a few decades ago but are now critical.

However, transitioning from a traditional job to a new, AI-oriented role isn't always straightforward. One of the most significant challenges is the skills gap. Many employees whose jobs are vulnerable to automation may not have the technical skills requisite for the new roles emerging in this AI-driven economy.

This skills gap poses a dilemma for businesses and governments alike. Organizations must invest in reskilling and upskilling their workforce. Governments, on the other hand, need to create educational initiatives that prepare the current and future workforce for a rapidly-evolving job market.

Remember when manufacturing jobs started moving overseas? The result was a significant disruption, but eventually, economies adapted. AI will inevitably cause similar disruptions, but the resilience of human ingenuity shouldn't be underestimated.

Interestingly, a report by McKinsey Global Institute suggests that while up to 375 million workers worldwide might need to switch job categories by 2030 due to AI, many new jobs will also be created. To put it bluntly, the human labor market is in a constant state of evolution. The challenge will be navigating this transition effectively.

Another fascinating aspect is the rise of the gig economy, partly fueled by AI. Platforms like Uber, which use sophisticated algorithms to match drivers with passengers, exemplify how AI can create new job opportunities. These might not be traditional full-time jobs, but they offer flexibility and are an essential part of the modern job landscape.

Then there's the question of social safety nets. If AI leads to significant job displacement, how do we support those who are affected? Universal Basic Income (UBI) is a concept that's been thrown around. Advocates argue that UBI could provide a financial cushion as workers transition to new roles. Whether or not UBI becomes a reality, it's clear that policy responses will need to adapt.

Industries that rely heavily on human intuition and creativity, like arts and culture, are likely to see less displacement. While AI can assist in creative processes, the human touch remains irreplaceable. Artists, designers, and writers may use AI as a tool to enhance their work rather than compete with it.

It's also crucial to look at job displacement through a global lens. Developed countries with robust education systems might adapt more quickly to these changes. In contrast, developing nations might struggle, exacerbating existing inequalities. Hence, international collaboration and knowledge sharing are essential to mitigate the negative impacts of AI on the workforce.

Job displacement doesn't just affect individuals but entire communities. Cities with economies based on industries likely to be automated will need to diversify. Take Detroit, for example, which faced severe economic turmoil when automobile manufacturing jobs declined. Forward-thinking city planning and investment in new industries can mitigate such risks.

Another critical factor is emotional resilience. Job loss can have significant psychological impacts, including stress and loss of identity. As a society, we need to foster an environment where people feel valued for their contribution, regardless of the nature of their work. Lifelong learning and adaptability should become core societal values.

It's a complex issue with no one-size-fits-all solution. Businesses, governments, educational institutions, and individuals all have a role to play in navigating this AI-driven era. Companies can lead by example, investing in their employees and being transparent about the impacts of AI. Governments can create policies that encourage innovation while protecting workers. Educational systems can evolve to teach skills that are relevant in an AI-centric world.

Looking ahead, we might see a symbiotic relationship between humans and machines rather than one overshadowing the other. AI isn't the end of human jobs; it's merely transforming them. With the right approach, we can turn job displacement into an opportunity for upward mobility and enhanced quality of life.

Every technological revolution brings its set of challenges and opportunities. This current wave of AI-driven transformation is no different. The key is to remain adaptable, inquisitive, and proactive. By doing so, we can ensure that the advancements in AI benefit as many people as possible, minimizing the negative impacts of job displacement and maximizing the array of new opportunities.

New Opportunities

The narrative around AI often zeroes in on job displacement, but let's dive into the opportunities that AI brings to the workforce. It's not just about what will be lost, but much more about what can be gained. The digital age has always been synonymous with change, and AI is no different. As we've seen in history, from the Industrial Revolution to the advent of the internet, every big leap in technology has brought with it a wave of new jobs, roles, and even industries.

Firstly, there's a burgeoning need for AI specialists. Companies across sectors are scrambling to hire data scientists, machine learning engineers, and AI ethicists. These roles are crucial for developing, implementing, and ensuring that AI systems perform ethically and efficiently. Not everyone needs to become an AI expert, of course, but the demand for these skills is skyrocketing. Training programs, online courses, and boot camps are popping up, making it easier than ever to acquire these competencies. This isn't just an opportunity for new graduates but also for mid-career professionals looking to pivot into more future-proof roles.

Moreover, AI is redefining existing jobs in ways that enhance human capabilities rather than replace them. Take customer service as an example. AI chatbots can handle routine inquiries, freeing up human agents to tackle more complex, emotionally nuanced interactions. Similarly, AI in healthcare doesn't negate the need for medical professionals but rather augments their ability to diagnose and treat patients by analyzing vast amounts of data faster and more accurately than a human ever could.

Another opportunity lies in the creation of hybrid roles—where AI and human skills intersect. Think about roles like AI-assisted marketing strategists, where marketers use predictive analytics to craft campaigns that are almost perfectly tailored to target audiences. Or financial advisors leveraging AI tools to provide investment recommendations backed by data-driven insights. These hybrid roles offer a unique blend of creativity and technical skills, opening up new career paths that didn't exist a decade ago.

Beyond individual roles, AI is also prompting the creation of entirely new industries. Autonomous vehicles are a prime example. As this technology advances, it is not just about creating the vehicles themselves but also building infrastructure, developing new forms of insurance, and even inventing forms of entertainment that cater to passengers rather than drivers. Each of these facets offers multiple job opportunities, from infrastructure planning to software development and more.

And let's not forget the startups. The AI revolution is fertile ground for entrepreneurial spirits. Startups focusing on everything from AI-driven personal health trackers to smart home systems are sprouting up everywhere. For those with a knack for innovation and a willingness to take risks, the AI landscape is wide open for disruption and advancement. Venture capital firms are increasingly interested in AI-focused ventures, providing the capital to transform groundbreaking ideas into successful businesses.

AI also promises to make the workplace more inclusive. Remote work, greatly enabled by AI tools that facilitate communication and collaboration, is here to stay. This shift has unprecedented potential to tap into a more diverse talent pool. It allows individuals who might be restricted by geography, physical ability, or personal circumstances to contribute meaningfully to the workforce. In many ways, AI can democratize access to work opportunities, breaking down traditional barriers and leveling the playing field.

Furthermore, consider the implications of AI in learning and development. With the aid of AI-driven educational platforms, employees can engage in lifelong learning tailored to their specific needs and career aspirations. Adaptive learning systems analyze individual performance and suggest personalized learning paths, making skill acquisition more efficient and engaging. This is profoundly transformative in a world where industries evolve rapidly, and skills become obsolete faster than ever before.

The creative arts are also not left out. AI tools are increasingly being used for content creation, from writing and music composition to film editing and even visual arts. While this might sound like a threat to traditional artists, it actually expands their toolkit, allowing them to push creative boundaries further than previously thought possible. Imagine filmmakers generating realistic CGI characters with just a few clicks or writers using AI to brainstorm and develop plot ideas.

Employers too stand to benefit enormously. AI can help in scouting talent by analyzing resumes for key skills and predicting a candidate's success based on past patterns. This not only speeds up the hiring process but also increases the likelihood of better matches between employers and employees. Additionally, AI-driven performance management systems can provide unbiased evaluations, thus fostering a fairer, merit-based work environment.

Then there's the gig economy—on steroids, thanks to AI. Platforms that match freelancers to short-term projects are now more sophisticated, making it easier for gig workers to find jobs that fit their skills and schedule. This flexibility allows for a more dynamic, responsive workforce. Imagine platforms that not only recommend gigs but also suggest ways to upskill to make oneself more marketable based on real-time demand forecasts. It's a new way of thinking about work, where boundaries and limitations are far less rigid.

Public sector jobs are also experiencing an AI-driven transformation. Governments and non-profit organizations can leverage AI to improve public services, from traffic management and crime prevention to social services distribution. These advancements make public sector roles more effective and rewarding, while also enhancing the quality of life for citizens. The ripple effect of AI in these areas can't be overstated; better-managed cities lead to more prosperous communities.

In conclusion, the advent of AI is a double-edged sword. While it does pose challenges, particularly in terms of job displacement, it equally, if not more potently, offers a panoply of new opportunities. The key to tapping into these opportunities lies in a forward-thinking approach—whether that means acquiring new skills, embracing hybrid roles, or venturing into AI-driven entrepreneurship. The landscape of work is changing, and those who adapt will find themselves at the vanguard of a new era, an era where human potential is amplified by artificial intelligence.

Chapter 9:
AI in Transportation

AI is dramatically reshaping the transportation industry, propelling us toward a future where autonomous vehicles and smart traffic systems redefine our daily commutes. This transformation is more than just a technological upgrade; it's a paradigm shift that promises safer, more efficient, and environmentally friendly travel. Imagine a world where your car is both your chauffeur and your co-pilot, skillfully navigating traffic while you focus on other tasks. AI-driven smart traffic systems, capable of predicting and alleviating congestion, are set to revolutionize urban mobility. These advancements offer not just incremental improvements but potentially radical changes in how we move from place to place, underlining the power and promise of AI to solve some of our most pressing transportation challenges. As we stand on the brink of this new era, it becomes crucial for business leaders, policymakers, and tech enthusiasts to understand and navigate these changes effectively.

Autonomous Vehicles

Autonomous vehicles aren't just a futuristic concept anymore; they're fast becoming a reality on roads around the world. Thanks to remarkable advances in AI and machine learning, self-driving cars are poised to revolutionize the way we think about transportation. Imagine a world where commuting isn't a daily grind but rather a seamless, efficient, and even enjoyable activity.

One might ask: what makes an autonomous vehicle truly "autonomous"? The answer lies in a rich tapestry of sensors, algorithms, and real-time data processing. These smart vehicles leverage technologies like LIDAR, radar, cameras, and GPS to gather information about their surroundings. Then, sophisticated AI algorithms process this data to make split-second driving decisions. It's a brilliantly orchestrated dance between hardware and software, designed to mimic human decision-making while minimizing risks.

Yet, the journey to achieve full autonomy has its challenges. One of the primary hurdles is dealing with unpredictable human behavior. While it's relatively straightforward for AI to navigate predefined routes or react to static objects, the dynamic nature of human-driven vehicles, pedestrians, and cyclists adds layers of complexity. Imagine an autonomous car at a busy intersection. It's not just about following traffic signals; it's about interpreting intentions, predicting movements, and ensuring passenger safety.

The potential benefits of autonomous vehicles are immense. For starters, they could drastically reduce traffic accidents, a significant number of which are caused by human error. Imagine a world where road fatalities are a thing of the past, and driving under the influence is no longer a concern. There's also the promise of enhanced mobility for those who can't drive, such as the elderly or disabled, offering a new level of independence.

However, integrating autonomous vehicles into existing transportation systems is no small feat. There's a need for robust infrastructure improvements, including smart traffic systems and updated road networks. Cities will have to adapt, incorporating new technologies into their urban landscapes. The shift won't just be technological but also sociopolitical, requiring collaboration between tech companies, governments, and the public.

The realm of autonomous trucking is particularly exciting. Long-haul trucking is a grueling job, often associated with driver fatigue, which can lead to accidents. Autonomous trucks could operate around the clock, improving efficiency and reducing costs. This shift could have a ripple effect across various industries, lowering transportation costs and, subsequently, the prices of goods.

But it's not all smooth sailing. The rise of autonomous vehicles also brings forth ethical considerations. For instance, how should a self-driving car prioritize lives in an unavoidable accident scenario? Should it protect its passengers at all costs or opt for the greater good? These are complex moral dilemmas that require thoughtful contemplation and clear regulatory frameworks.

Moreover, there's the question of cybersecurity. Autonomous vehicles are, in essence, computers on wheels. As such, they are susceptible to hacking and cyber-attacks. Ensuring that these vehicles are secure is paramount to prevent potentially catastrophic outcomes.

From an economic standpoint, autonomous vehicles could reshape job markets. There will be a decline in demand for traditional driving jobs, but new opportunities will emerge in tech, vehicle maintenance, and infrastructure development. The key will be to prepare the workforce for this shift, investing in education and training programs to equip workers with the necessary skills.

On the consumer side, the transition to autonomous vehicles will be gradual. Early adopters may be tech enthusiasts eager to embrace new innovations, but mass adoption will hinge on trust and affordability. Manufacturers and policymakers will need to address public concerns, demonstrating that these vehicles are not only safe but also beneficial for society as a whole.

Autonomous vehicles also have the potential to reshape urban planning and real estate. With fewer parking spaces required and a decrease in traffic congestion, city landscapes could transform. Public spaces might expand, and the real estate market could shift as people move away from densely populated areas, no longer tethered to traditional commuting routes.

As we stand on the cusp of this transportation revolution, it's crucial to look ahead and anticipate the broader societal impacts. The environmental benefits, for instance, are promising. Autonomous vehicles can operate more efficiently than human drivers, optimizing routes and reducing fuel consumption. This shift could play a vital role in efforts to combat climate change.

Furthermore, the integration of autonomous vehicles with ride-sharing services could bring about a future where personal car ownership is less common. Instead, mobility as a service (MaaS) could become the norm, with people opting for on-demand transportation solutions. This paradigm shift could lead to a more sustainable and efficient use of resources.

The collaboration between automotive giants, tech companies, and startups is already accelerating the development of autonomous vehicles. Companies like Tesla, Waymo, and Uber are at the forefront, constantly pushing the boundaries. Their efforts are complemented by academic research and governmental initiatives, fostering an ecosystem where innovation thrives.

However, as we navigate this journey, it's essential to maintain a balance between innovation and regulation. Policymakers must craft laws that foster innovation while ensuring public safety. Open dialogues between stakeholders, including the public, are crucial in shaping a future that benefits everyone.

In conclusion, the rise of autonomous vehicles marks a pivotal moment in the history of transportation. From enhancing safety to transforming urban landscapes and offering new economic opportunities, the potential is vast. But with this promise comes responsibility. It's up to innovators, regulators, and society to steer this technology in a direction that prioritizes ethics, safety, and inclusiveness, ensuring a brighter future for all.

Smart Traffic Systems

Imagine a world where traffic jams are a relic of the past, where your daily commute is as smooth as a lazy Sunday drive, and emergency vehicles navigate through city streets as if the roads were tailor-made for them. Smart traffic systems, driven by the power of AI, promise to turn this fantasy into reality. They are the unseen force working tirelessly behind the scenes to create a more efficient, sustainable, and safer transportation experience for all.

The concept of smart traffic systems revolves around using AI to analyze and optimize traffic flow in real-time. Unlike traditional traffic management, which relies heavily on fixed schedules and outdated data, AI-based systems dynamically adjust to current conditions. By leveraging machine learning algorithms, these systems can predict traffic patterns, prevent bottlenecks, and even reduce the risk of accidents.

One of the most compelling aspects of smart traffic systems is their ability to manage traffic lights intelligently. Using data from various sources—such as cameras, sensors, and even weather forecasts—AI can determine the optimal timing for traffic lights. This way, traffic doesn't just flow; it dances. Congestion becomes manageable, fuel consumption lowers, and the environment benefits from reduced emissions.

Micromanagement at Scale

For those who love efficiency, consider how AI micromanages traffic at the scale of an entire city. Let's say, for example, an accident happens on a major highway. A smart traffic system would immediately reroute vehicles through less congested streets, communicate with other vehicles about the delay, and adjust traffic signals to avoid secondary jams. It's a symphony of coordinated actions that happen without drivers even noticing.

What's fascinating is how these systems continually learn and improve. Each day brings new data points, which feed into machine learning algorithms, making the entire system smarter over time. Severe weather conditions, a public event, or an unexpected road closure: the longer the system operates, the better it becomes at handling such unpredictable variables.

Data Fusion

However, to understand the true potential of smart traffic systems, one should grasp the concept of data fusion. This means integrating data from multiple sources to provide a comprehensive view of traffic conditions. Cameras, radar sensors, GPS data from vehicles, and even social media updates can provide valuable information. AI algorithms fuse this data to create actionable insights, transforming raw information into a well-orchestrated traffic control strategy.

Let's not forget about the bus fleets, trains, and bicycles that share the urban landscape with cars. Smart traffic systems are also equipped to harmonize these different modes of transport. During peak hours, AI can prioritize public transport vehicles or create dedicated lanes for them, ensuring that everyone reaches their destination more efficiently.

Emergency Response

In life-threatening emergencies, every second count. Smart traffic systems can be lifesavers—literally. When an emergency vehicle needs to get through congested traffic, AI can clear its path by dynamically adjusting traffic lights and rerouting other vehicles. Imagine a scenario where an ambulance doesn't have to worry about gridlocks; the system ensures it reaches the hospital as quickly as possible with minimal delays.

These capabilities are not just theoretical. Cities like Los Angeles and Amsterdam have already begun implementing AI-driven traffic management systems. Early results have been promising: reduced travel times, lower emissions, and higher compliance with traffic rules. While the technology is still evolving, its potential to transform urban mobility is undeniable.

Public Sentiment and Challenges

But let's address the elephant in the room: public sentiment and challenges. Introducing AI into something as daily and ubiquitous as commuting isn't without its hurdles. People may be skeptical about data privacy, particularly when personal vehicle data is involved. It's crucial for policymakers and tech developers to build trust. Rigorous data security measures and transparent policies can go a long way in allaying fears.

There's also the financial aspect. Upgrading a city's traffic management system to a smart one isn't cheap. The installation of sensors, maintaining the infrastructure, updating software—these are significant investments. However, when balanced against the long-term benefits like reduced fuel consumption, improved air quality, and enhanced road safety, the cost seems justifiable.

The Bigger Picture

Zooming out a bit, smart traffic systems are part of the larger puzzle that is smart cities. They connect seamlessly with other city services like energy management, waste collection, and even crime monitoring. In many ways, efficient traffic management becomes a stepping stone for larger, more ambitious urban innovations. The smarter our cities become, the better equipped we are to handle the complexities of modern life.

Looking ahead, the integration of quantum computing could take smart traffic systems to an entirely new level. Current algorithms, while highly efficient, have their limitations. Quantum computing could provide even faster data processing capabilities, enabling real-time adjustments on a scale previously thought impossible.

Inspirational Possibilities

Think about the inspiration behind all this innovation: the idea that technology should serve humanity, making life not just more comfortable, but also more egalitarian. When you're not stuck in traffic, when the air is cleaner, when emergency services can save lives faster, the quality of life across all societal layers improves. This isn't just about getting from point A to point B; it's about creating a future where technology and human well-being go hand-in-hand.

There's a saying, "The journey of a thousand miles begins with a single step." Smart traffic systems are that crucial first step towards creating a brighter, more equitable future. They embody what AI can achieve when applied thoughtfully and inclusively.

So, as we steer into this new era of AI-driven innovation, let's keep our eyes on the road ahead, not just for the technological marvel it represents, but for the extraordinary possibilities it unlocks for all of us.

With great power comes great responsibility. As tech enthusiasts, business leaders, and advocates for a smarter future, it's up to us to ensure that smart traffic systems are implemented with care, precision, and a commitment to improving life for everyone. It's an exciting journey, filled with opportunities to rethink, reshape, and redefine the way we move through our world.

Chapter 10:
AI in Communication

In today's digitally connected world, AI is revolutionizing the way we communicate, making interactions smoother and more intuitive than ever before. Imagine voice assistants that not only understand but also predict your needs, or natural language processing tools that break down language barriers, bringing people closer across the globe. Businesses are harnessing this power to enhance customer service, streamline operations, and even personalize marketing in ways previously unimagined. But it's not just about efficiency; AI in communication fosters richer, more meaningful engagements, turning data streams into dialogues that feel almost human. As we explore these advancements, one thing becomes clear: AI is not just a tool but an enabler of a connected, collaborative future.

Natural Language Processing

It's astonishing how far we've come in teaching machines to understand and generate human language. Natural Language Processing, or NLP for short, sits at the confluence of computer science, artificial intelligence, and linguistics. This technology is rapidly evolving and plays an integral role in transforming the way we communicate with machines and one another. NLP isn't just about translating text from one language to another; it's about teaching computers to understand the nuances, idioms, and even the emotions embedded in our words.

Think about how often we interact with digital assistants like Siri, Alexa, or Google Assistant. These interactions are made possible by sophisticated NLP algorithms that parse our spoken words, interpret their meaning, and provide relevant responses. They leverage vast amounts of language data to make sense of context and intent, learning continually to become more accurate and natural in their understanding. The applications are vast and varied, ranging from simple query answering to complex tasks like drafting emails, summarizing articles, and composing music lyrics.

Natural Language Processing isn't a one-size-fits-all technology. It encompasses a broad spectrum of techniques designed for various challenges. Tokenization breaks down text into smaller chunks, enabling easier analysis. Named Entity Recognition (NER) identifies and classifies key components within text, such as names of people, companies, or locations. Sentiment analysis gauges emotional tone, deciphering whether the language used is positive, negative, or neutral. Each technique addresses specific aspects of language, making the whole system more adaptive and versatile.

Consider machine translation, one of the most impressive feats of NLP. In the past, translating a document from English to Chinese would have required a human translator with bilingual expertise. Today, neural machine translation models, powered by deep learning, can translate entire articles almost instantaneously. While not perfect, these systems continue to improve, reducing the language barrier that has long hampered global communication.

The realm of text generation is equally fascinating. GPT-3, a language model developed by OpenAI, can produce coherent and contextually accurate text based on a given prompt. Capable of writing essays, poems, and even code, GPT-3 demonstrates the unprecedented potential of NLP in content creation. It can assist writers with ideation, automate routine writing tasks, and generate personalized content, making for a much more efficient creative process.

Certainly, NLP goes beyond just text. Voice processing is another critical discipline under its umbrella. Voice recognition technologies use NLP to transcribe spoken words into text, which can then be analyzed or acted upon. Companies are investing heavily in voice interfaces, recognizing the convenience they offer over traditional input methods. Imagine a world where typing is a relic of the past, replaced by seamless voice commands that understand multiple languages, dialects, and accents.

Another exciting application is sentiment analysis, which is particularly valuable in business contexts. By analyzing public reviews, social media posts, and customer feedback, companies can gauge consumer sentiment in real-time. This invaluable data helps businesses refine their products, tailor their marketing strategies, and improve customer service. It's a direct line to the consumer's mind, enabling brands to be more responsive and customer-centric.

While the impact of NLP on businesses and technology is significant, its potential is boundless when applied to social good. Take, for example, its role in healthcare. NLP can parse clinical notes, helping doctors sift through patient records to find crucial pieces of information quickly. This speeds up diagnoses and ensures that critical details aren't overlooked. In mental health, sentiment analysis can identify patterns in patients' language indicative of their emotional state, aiding in early intervention and treatment.

The educational sector also stands to benefit immensely. NLP-driven applications can provide personalized learning experiences by analyzing students' reading and writing. Intelligent tutoring systems offer real-time feedback, guiding students through complex subjects at their own pace. This technology promises to make education more inclusive and effective, accommodating diverse learning styles and needs.

Innovation in NLP isn't without its challenges. Language is incredibly complex, rich with idioms, metaphors, and sarcasm that even humans sometimes struggle to interpret. Ambiguities in language can lead to misunderstandings, and ensuring that NLP systems are inclusive and unbiased is a significant hurdle. Developers and researchers must diligently work to minimize the risks of bias in training data, lest these systems propagate or even amplify societal inequities.

Ethics in NLP development is a hot topic. With great power comes great responsibility, and those at the helm of NLP innovation must navigate this landscape carefully. Ensuring transparency, accountability, and fairness in these systems is crucial. The decisions made now will shape the future of how humans and machines interact—responsible development will ensure that this future is equitable and beneficial for all.

The road ahead for NLP is both exciting and challenging. As we continue to push the boundaries, the line between human and machine communication becomes increasingly blurred. The next frontier may see machines understanding the subtlest nuances of human emotion, irony, and context. As businesses, educators, and technologists invest in these advancements, the potential for NLP to transform our world is limitless.

Ultimately, NLP is a testament to the incredible progress we've made in computational linguistics and AI. Its applications are diverse, affecting numerous aspects of our daily lives and opening up new possibilities in how we interact with information and each other. Moving forward, the goal remains to make these systems more human-like in their understanding while ensuring they are used ethically and responsibly. Achieving this balance will empower us to harness the true potential of Natural Language Processing, driving innovation and enhancing human communication in unprecedented ways.

Voice Assistants

Voice assistants have swiftly integrated into our daily lives, transforming the way we interact with technology. Unlike traditional interfaces requiring touch and type, voice assistants offer a more natural and seamless way to access information and perform tasks. It's almost magical – we speak, and they respond, executing commands with remarkable precision. This section delves into the multifaceted role of voice assistants, their technological underpinnings, and their broader impact on communication.

The rise of voice assistants began modestly but has escalated into a tech phenomenon. Apple's Siri, introduced in 2011, was among the pioneers, followed by Google Assistant, Amazon Alexa, and Microsoft's Cortana. These digital companions combine natural language processing (NLP) and machine learning to decode and process human speech, convert it into actionable commands, and provide responses. The seamless interaction they afford is revolutionizing our relationship with machines.

Voice assistants are grounded in advanced natural language processing techniques. They discern user intent through context, syntax, and semantics, a feat made possible by sophisticated algorithms and vast data sets. By analyzing massive volumes of spoken and written language, these systems learn to understand the nuances of human speech, including accents, slang, and varying intonations. Several iterations and continuous learning cycles allow voice assistants to improve their accuracy and responsiveness.

One can't ignore the convenience voice assistants bring into our lives. Whether it's setting a reminder, playing a favorite song, controlling smart home devices, or retrieving instant news updates, voice assistants have become indispensable. They streamline mundane tasks, enabling us to focus on more critical activities, thus enhancing productivity. The hands-free interaction they offer is particularly beneficial in scenarios where multitasking is essential, such as driving or cooking.

However, the application of voice assistants goes beyond mere convenience. In the business sector, they are transforming customer service and operational efficiency. Companies leverage voice assistants for handling customer inquiries, processing orders, and even providing guided assistance. Chatbots integrated with voice capabilities can manage high volumes of customer interactions concurrently, offering timely support and freeing up human agents for more complex issues. This not only optimizes resource allocation but also significantly enhances customer satisfaction.

Moreover, voice assistants are finding their place in educational and healthcare contexts. In education, they can serve as personal tutors or learning aids, helping students access information quickly and seamlessly. In healthcare, assistants like those developed by Babylon Health assist in symptom checking and patient triage, providing preliminary diagnoses and advice, which can be a lifeline in remote or underserved areas.

Privacy concerns cannot be overlooked when discussing voice assistants. Given the always-on nature of these devices, they continuously listen for wake words, raising questions about data security and unauthorized access. It's crucial for developers to ensure robust encryption protocols and transparent data handling practices to safeguard user privacy. Regulations like GDPR in Europe are steps toward providing such protection, ensuring data collected via voice assistants is handled ethically and securely.

Voice assistants also herald a paradigm shift in accessibility. For the differently-abled, particularly those with visual or motor impairments, these tools open up new avenues for interaction with technology. By bypassing the need for physical input devices, voice assistants empower users to engage with digital content and smart devices, fostering greater independence and inclusion.

Expanding on the technological aspects, voice recognition technology at the core of these assistants utilizes neural networks, emulating the human brain's structure. These networks are trained on diverse datasets to fine-tune their understanding of language subtleties. Continuous feedback loops and user interactions present real-world scenarios, allowing these models to learn and adapt dynamically.

The potential for future development in voice assistants is tremendous. Integration with Internet of Things (IoT) will enable even broader applications, from fully automated smart homes to advanced industrial systems. Imagine a future where voice assistants manage a whole city's electrical grid, optimizing energy consumption and pinpointing issues instantaneously. The blending of AI with voice recognition technology could be the key to unlocking such futuristic scenarios.

The competitive landscape among tech giants has spurred rapid innovation in the voice assistant arena. Companies continually strive to outdo each other, leading to regular updates and new features. This competition fosters an environment of continuous improvement, benefiting end-users with more sophisticated and capable assistants.

Voice assistants are also becoming culturally adaptive, with companies working to infuse local languages, dialects, and cultural nuances into these systems. This localization is essential for global acceptance, enabling non-English speaking populations to leverage the benefits of voice-activated technologies. By doing so, voice assistants can achieve a truly global reach.

Looking toward the future, the integration of emotion detection in voice assistants is an exciting frontier. By recognizing emotional cues in speech, these systems can provide more empathetic and contextually appropriate responses. This development would mark a significant leap, moving from mere functional assistance to emotionally intelligent interactions that can provide comfort and support in more meaningful ways.

Voice assistants stand at the intersection of convenience, efficiency, and innovation. Their evolution from simple command-followers to sophisticated conversational agents epitomizes the strides made in AI and natural language processing. As these systems become more integrated into our lives, their impact on communication will only grow, continuing to shape and redefine our interactions with technology. The future is not just about having smarter assistants but about fostering smarter lives, where technology seamlessly supports and enhances human potential.

Chapter 11:
AI and Entertainment

As we dive into the electrifying domain of AI and entertainment, it's impossible not to be wowed by the transformative power of algorithms that are shaping our daily experiences with content. Imagine logging into your favorite streaming service, and within seconds, a tailored array of shows and movies appears—courtesy of sophisticated AI-driven personalized recommendations that seem to read your mind. Then there's content creation. AI-generated scripts, music, and even animations are no longer science fiction but a vivid reality, pushing creative boundaries and challenging conventional norms. While human intuition and creativity aren't going anywhere, AI is becoming a potent collaborator, offering tools that amplify and extend our artistic capabilities. The entertainment industry, always a playground for the imagination, is evolving faster than ever. Remember, we're not just passive consumers of AI; we're co-authors in a narrative that's unfolding with dizzying speed and infinite possibilities.

Personalized Recommendations

Imagine you're sitting at home after a long day, ready to unwind with your favorite streaming service. Without even thinking, you trust that the platform will suggest content you'll enjoy. This seemingly simple interaction is the culmination of complex AI algorithms designed to understand you better than you may understand yourself. Personalized recommendations have revolutionized the way we consume media, and they're just the beginning.

At its core, the concept is not new. For decades, businesses have relied on recommendations to drive engagement and sales. What's changed is the precision and scale enabled by artificial intelligence. Modern AI employs machine learning techniques that analyze vast amounts of data to predict what you might like next. These algorithms don't just consider your past behavior; they also look at broader patterns and trends.

Take Netflix, for example. This entertainment giant utilizes a layered recommendation engine. Initially, it was based solely on user ratings. Now, it evaluates viewing history, genre preferences, and even the time of day you're most likely to watch certain types of content. All of these data points combine to create a more personalized experience.

But how do these systems work on a technical level? Generally, they rely on two key techniques: collaborative filtering and content-based filtering. Collaborative filtering looks at user behavior and identifies patterns. For instance, if you and another user have shared interests, the algorithm will suggest movies or shows that the other person has watched, assuming you might enjoy them too. On the other hand, content-based filtering analyzes the attributes of the content itself, suggesting items with similar characteristics to what you've already liked.

The beauty of personalized recommendations lies in their constant evolution. AI systems get smarter over time, adapting to your changing tastes. For instance, if you suddenly develop an interest in documentaries, your recommendations will start reflecting this new preference in no time. This fluidity is achieved through continuous learning algorithms that process real-time data.

However, it's not just entertainment platforms that benefit from this. E-commerce sites like Amazon employ sophisticated recommendation algorithms to drive sales. These systems analyze your shopping habits, viewing history, and even your browsing patterns to suggest products you might find appealing. The recommendation engine is so advanced that it often feels like the platform knows what you want before you do.

In the world of music, Spotify has set a standard with its use of AI for personalized recommendations. Utilizing both collaborative and content-based filtering, along with natural language processing, Spotify analyzes millions of songs and user behaviors to curate playlists tailored to individual tastes. The result is a highly personalized experience that seems almost magical in its accuracy.

Despite the many advantages, some concerns surround the use of AI for personalized recommendations. One major issue is the creation of filter bubbles, where users are only exposed to content that reinforces their existing views and interests. This can limit discovery and stifle diversity. Another concern is data privacy. These systems rely on gathering and analyzing massive amounts of personal data, raising questions about how this data is stored, used, and protected.

Businesses leveraging these technologies must ensure they strike a balance between personalization and privacy. Transparency with users about how their data is collected and used is crucial. Additionally, incorporating mechanisms for users to have some control over their recommendations can mitigate the risks of filter bubbles.

Incorporating AI into personalized recommendations also carries significant potential for enhancing user experience beyond just entertainment and shopping. Educational platforms can use AI to recommend courses and resources tailored to individual learning styles and progress. Fitness apps can suggest personalized workout routines based on user data. The possibilities are almost endless.

As we look to the future, the integration of AI in personalized recommendations will only grow more sophisticated. Emerging trends like deep learning and neural networks are pushing the boundaries of what's possible. These technologies enable algorithms to understand complex patterns and make even more accurate predictions. Imagine a future where your smart home system knows precisely what you need, from suggesting the next book for your digital library to adjusting your home's lighting based on your mood.

There's also the impending influence of quantum computing, which promises to supercharge AI capabilities. Quantum algorithms could process and analyze data at unprecedented speeds, making personalized recommendations even more effective and near-instantaneous. While this is still on the horizon, the potential impact can't be understated.

The implications for businesses are significant. Those that master personalized recommendations can achieve a competitive edge, driving user engagement, loyalty, and ultimately revenue. However, this requires a deep understanding of AI technologies and a commitment to continual innovation.

In summary, personalized recommendations are more than just a feature; they represent a fundamental shift in how businesses interact with their customers. Through leveraging advanced AI techniques, companies can create highly individualized experiences that resonate on a personal level. While challenges like data privacy and filter bubbles need addressing, the opportunities far outweigh these concerns. As AI continues to evolve, so too will the capabilities and applications of personalized recommendations, making them an integral part of our digital lives.

Content Creation

AI has revolutionized many sectors, but its impact on content creation is particularly noteworthy. Imagine a world where movies, music, and literature are not just consumed faster but are created at an unprecedented pace. With AI, creativity and technology intersect in ways we've only seen in science fiction until now. AI-driven content creation is reconfiguring the entertainment industry, both in how content is made and how it is experienced by audiences. So, let's dive in.

In traditional media, creating high-quality content often takes months, if not years. Whether it's a feature film or a best-selling novel, the creative process involves brainstorming, scripting, editing, and finally producing. AI disrupts this cycle by speeding up various stages of production. For instance, scriptwriting software powered by AI can draft dialogue or plot points, saving screenwriters countless hours. This isn't just theoretical; tools like ScriptBook use machine learning algorithms to predict box office success based on script quality, enabling studios to make informed decisions before investing in a project.

When it comes to filmmaking, AI isn't just stopping at the script. Directors are making use of AI for storyboarding, scene planning, and even selecting the best takes during editing. AI-powered software can analyze hours of footage and pick the most visually appealing shots, greatly alleviating the editor's workload. AI tools like SceneSkim enhance this process, making it more efficient while preserving the director's creative vision. And it's not just behind the scenes; AI is impacting the casting process as well. AI-driven platforms can analyze an actor's previous performances to predict their suitability for a new role.

Music is another area where AI is leaving an indelible mark. AI-generated music is no longer a mere curiosity; it's becoming mainstream. Platforms like OpenAI's Jukedeck and Google's Magenta allow musicians to create new compositions effortlessly. These AI tools can generate melodies, chords, and even complex symphonies based on given inputs. The boundary between human and machine-generated music continues to blur as AI composes works that are nearly indistinguishable from those created by human musicians.

Moreover, AI has proven itself invaluable in music production and mixing. AI algorithms like LANDR can master tracks by analyzing and optimizing audio elements, bringing out the best aspects of a song. This democratizes music production, enabling independent artists to achieve studio-quality sound without needing expensive equipment or expertise. Suddenly, anyone with a laptop and an internet connection can produce professional-grade music, opening doors to untapped talent across the globe.

Literature, too, is being transformed. AI can assist writers in several ways, ranging from simple grammar checks to more complex tasks like generating text based on a given style or tone. Projects like OpenAI's GPT-3 have demonstrated the ability to create coherent, engaging narratives on a wide range of topics. Authors can use these tools as a writing assistant or even as a collaborator, brainstorming ideas or developing plots. The rise of AI doesn't threaten the domain of traditional authorship but augments it, offering new ways to stretch the boundaries of creative writing.

AI tools are also shaking up journalism. Automated journalism systems can generate news articles from factual data within seconds. Companies like the Associated Press and Bloomberg are already leveraging AI to cover financial earnings reports, sports events, and even natural disasters. These tools analyze data and create text-based summaries almost instantaneously, ensuring that news dissemination is more immediate and accurate. This allows human journalists to focus more on in-depth reporting and less on routine tasks.

It's essential to address the ethical considerations surrounding AI-driven content creation. The integration of AI in creative fields raises questions about the originality and ownership of work. If an algorithm writes a song or a novel, who holds the copyright? Moreover, concerns around biases in AI models can't be ignored. AI systems learn from existing data, which may reflect societal prejudices. Ensuring that AI tools promote diversity and ethical standards in content creation is crucial for the technology's responsible adoption.

Collaboration between artists and AI is becoming the norm rather than the exception. The creation of digital art has seen exponential growth, with artists using AI to generate unique pieces that might be impossible to create manually. Generative Adversarial Networks (GANs) are at the forefront of this revolution. They consist of two neural networks contesting each other, producing art that challenges the viewer's sense of what is real and what is machine-made. This partnership between human intuition and machine precision is crafting a new frontier in visual arts.

When we talk about video games, AI's contribution is multifaceted. It's not just about improving graphics or gameplay; AI is now involved in creating entire worlds and narratives. Procedural content generation, driven by AI, can produce expansive game worlds that evolve and adapt based on player interaction. Imagine a game where the storyline shifts dynamically, offering a unique experience every time you play. AI algorithms tailor challenges and puzzles to individual skill levels, ensuring that the gaming experience is both engaging and rewarding.

Narrative-driven games are leveraging AI to create more profound emotional experiences. Non-Player Characters (NPCs) powered by AI exhibit realistic behaviors and dialogues, making interactions feel more authentic. AI can analyze player choices and adapt the storyline, providing a personalized narrative arc. These advancements are not just enhancing player engagement but are also broadening the horizons of what video games can achieve as an art form.

As we delve into user-generated content, platforms like TikTok and YouTube are utilizing AI in fascinating ways. Algorithms curate and recommend personalized content, enhancing the user experience by delivering what you didn't know you wanted to watch. These platforms also employ AI for automatic tagging, content moderation, and even video editing. Ever wonder how some videos seem perfectly cut to the beat of the music? That's AI at work. Such advancements make it easier for creators to produce compelling content without extensive technical skills.

The transformative power of AI in content creation is also being recognized by streaming giants like Netflix and Spotify. These companies use machine learning algorithms to analyze viewing and listening habits, offering recommendations that feel almost eerily accurate. But they're going a step further. Netflix, for instance, is experimenting with AI to create customizable content. Imagine watching a show where the ending changes based on your preferences or previous viewing history. AI is making personalized storytelling a reality, evolving the passive viewing experience into an interactive journey.

The future of AI in content creation offers tantalizing possibilities. AI could enable real-time collaboration between artists across the globe, breaking down geographic and linguistic barriers. We might see virtual reality experiences generated on the fly, tailored to individual user inputs, making each experience unique. As AI continues to evolve, the creative landscape will become increasingly democratized, making high-quality tools accessible to anyone with a vision.

In summary, AI is reshaping the content creation landscape in profound ways. From scriptwriting and music composition to gaming and digital art, AI is not merely a tool but a collaborator in the creative process. While ethical considerations and challenges persist, the promise of AI in unleashing human creativity is undeniable. As we stand on the cusp of this new era

Chapter 12:
The Future of AI in Education

As we stand on the brink of a new era in education, AI isn't just a fleeting trend but a transformative force ready to elevate how we teach and learn. Imagine a world where adaptive learning systems meticulously tailor the educational experience to each student's needs, pacing lessons to match individual progress and uncovering potential at an unprecedented scale. Virtual tutors, far from being mere assistants, become personal mentors available around the clock, breaking down complex concepts into digestible insights. This brave new landscape redefines accessibility, bringing top-tier education to even the most remote corners of the globe. The classroom of the future is a dynamic ecosystem where AI-powered tools don't replace educators but augment their capabilities, allowing them to focus on nurturing creativity and critical thinking. By merging human intuition with machine precision, we unlock a future where every student's potential is not just recognized but fully realized.

Adaptive Learning Systems

Imagine walking into a classroom where every student's learning experience is meticulously tailored to their specific needs, abilities, and even their moods. That's not the distant future; it's the kind of present we're gradually stepping into with the advent of adaptive learning systems. These systems are revolutionizing the way we think about education by leveraging data and predictive analytics to create highly individualized learning experiences.

At its core, an adaptive learning system uses AI algorithms to assess a student's current level of understanding and adapt the educational content accordingly. Think of it like a really smart tutor who knows exactly what you need to learn next, and how best to present that information to you. This isn't just about filling gaps in knowledge; it's about optimizing the entire learning journey for each individual.

Consider the possibilities. A student struggling with algebra might receive additional practice problems tailored to pinpoint their weaknesses, while another student who excels in the same subject could be given more complex challenges to prevent boredom. The system can analyze patterns in a student's performance, offering real-time feedback and modifying the difficulty level on the fly. It's not about making education easier; it's about making it more efficient and effective.

Many of us have had the experience of being lost in a classroom, unable to keep up with the teacher's pace. Adaptive learning systems can alleviate this pain point by ensuring that no student gets left behind. Conversely, it also means that faster learners are not held back by a one-size-fits-all curriculum. The adaptability of these systems creates a more balanced and inclusive educational environment.

The beauty of adaptive learning isn't just in its capability to customize learning experiences; it's also in its ability to scale. Traditional education methods often struggle with large class sizes, where individualized attention from the teacher is a rare luxury. Adaptive learning technologies can handle hundreds, even thousands, of learners at the same time, each receiving a personalized education pathway.

Machine learning and data analytics play a critical role here. These technologies enable the system to constantly learn and improve from the data it collects. Over time, the system becomes better at predicting which methods work best for different types of learners. This ongoing cycle of improvement and adaptation makes these systems incredibly powerful tools for educators.

Let's not forget the role of engagement. Traditional methods often fail to capture the interest of all students, leading to disengagement and dropout rates. Adaptive learning can incorporate elements of gamification, interactive content, and even artificial intelligence-driven avatars to make learning more engaging. For example, a student might receive rewards for completing a series of challenging tasks, or they might interact with a virtual tutor that can provide instant feedback.

One might wonder, with such heavy reliance on technology, what becomes of the human educator? Far from being replaced, teachers are transformed into facilitators and guides within this new educational paradigm. Freed from the constraints of delivering the same lecture to every student, teachers can now focus on more meaningful interactions, providing guidance and support where it's most needed. This shift allows educators to develop deeper relationships with their students, acting more like coaches than lecturers.

Furthermore, teachers can use data analytics provided by adaptive learning systems to monitor student progress and intervene when necessary. This data-driven approach allows for more informed decisions, enabling teachers to target their efforts more effectively. Instead of guessing which students might need extra help, educators can rely on concrete data to pinpoint issues quickly.

The potential impact on lifelong learning is also significant. Adaptive learning systems can extend beyond K-12 and higher education to professional development and remote learning environments. In an era where continuous skill development is crucial to career advancement, these systems can offer tailored training programs that adapt to the learner's pace and existing knowledge base. Imagine a world where reskilling and upskilling are seamlessly integrated into our daily lives, making education a continuous, adaptive journey.

There are, of course, challenges and considerations that come with implementing adaptive learning systems. Privacy and data security remain paramount concerns. Collecting vast amounts of data on students raises questions about how that information is stored, who has access to it, and how it can be protected. Ethical considerations also come into play, ensuring that these systems are fair and do not perpetuate existing inequalities or biases.

Another challenge is the digital divide. Not all students have equal access to the technology required for adaptive learning systems. Bridging this gap will be crucial for the widespread adoption and success of these technologies. Schools and institutions will need to invest in infrastructure and provide necessary resources to ensure all students can benefit equally.

Nonetheless, the strides being made are truly inspirational. Innovative companies and institutions are piloting these systems with promising results. Some schools have reported significant improvements in student engagement and academic performance. Governments and educational authorities are beginning to recognize the potential and are investing in research and development to bring these adaptive systems to more classrooms.

The journey of building an adaptive learning system is akin to the process of learning itself - iterative, continuous, and constantly evolving. The ultimate goal is not just to enhance the learning experience but to democratize education. By making high-quality, personalized education accessible to everyone, we can create a more equitable and informed society. Imagine a world where every student, regardless of their background or circumstances, has the opportunity to reach their full potential.

In closing, adaptive learning systems represent a seismic shift in the education landscape. They offer a glimpse into a future where learning is not confined by traditional boundaries and where each individual can enjoy a customized educational journey. The key will be to navigate the challenges thoughtfully and ethically, ensuring that we harness the incredible potential of these technologies to benefit all learners. This evolution in education will not only shape the future of AI but also redefine how we approach the art of learning itself.

As we embrace these innovations, it's clear that the future of education is incredibly bright. Adaptive learning systems stand as a testament to the extraordinary possibilities that arise when technology and human ingenuity intersect. The next chapter in education has already begun, and it's one that promises to unlock new avenues of opportunity for learners around the globe.

Virtual Tutors

The landscape of education is evolving at a pace we can hardly keep up with, and one of the key drivers of this change is Artificial Intelligence (AI). More specifically, virtual tutors are gradually becoming a cornerstone in the educational ecosystem. Unlike traditional tutors, who require physical presence and adherence to a schedule, virtual tutors offer a level of flexibility and accessibility that is unprecedented. Imagine having a tutor available 24/7, prepared to assist you with any subject matter, no matter how obscure or complex.

On the surface, virtual tutors might seem like a simple extension of the existing e-learning platforms, but they are so much more. Powered by sophisticated algorithms, these AI-driven educators can adapt their teaching methods based on the learner's unique needs. They can identify gaps in understanding and tailor lessons to fill those gaps, ensuring that no student is left behind. This personalized approach can make learning more engaging and effective, leading to better outcomes.

One of the most intriguing aspects of virtual tutors is their ability to learn from their interactions. Through machine learning, these systems continually improve their teaching strategies. The more they interact with students, the better they become at identifying common pitfalls and misconceptions. This ongoing learning process enables virtual tutors to provide increasingly accurate and helpful guidance.

The implications of virtual tutors go beyond just personalized learning. They could bridge the gap between traditional education and the growing need for lifelong learning. In today's fast-paced world, the ability to continuously learn and adapt is crucial. Virtual tutors can assist adults in acquiring new skills and knowledge, making it easier for them to pivot in their careers or adapt to new technological advancements.

Let's not forget the potential impact on underprivileged communities. In many parts of the world, quality education is a luxury that few can afford. Virtual tutors, with their low-cost and scalable solutions, could democratize education. Imagine a child in a remote village having access to the same quality of education as a student in a city. The playing field could be leveled in a way we've only dreamed about.

However, it's not all sunshine and rainbows. There are challenges and ethical considerations that come with the integration of virtual tutors into our education systems. One major concern is the potential for these systems to perpetuate existing biases. If the data used to train these tutors is biased, the lessons they provide will also be biased. Ensuring that the algorithms are fair and impartial is a hurdle that needs to be addressed.

Moreover, the human touch in education is something that technology might never fully replicate. Teachers do more than just impart knowledge; they mentor, inspire, and foster a sense of community. Virtual tutors, no matter how advanced, may not be able to fill this emotional and social role. This raises questions about the balance between technology and human interaction in education.

Businesses and tech leaders need to think critically about how to implement these systems responsibly. It's essential to involve educators in the design and deployment of virtual tutors. Their insights can help ensure that these tools are used to complement, rather than replace, the vital role of human teachers.

Despite these challenges, the potential benefits of virtual tutors are too significant to ignore. They offer a glimpse into a future where education is more tailored, accessible, and equitable. The technology is still in its nascent stages, but the rapid advancements in AI suggest that we are on the brink of an educational revolution.

For tech enthusiasts, the development and refinement of virtual tutors present exciting opportunities. From natural language processing to data analytics, the technological innovations required to make virtual tutors effective are at the cutting edge of AI research. This is a field ripe for innovation and investment, with the potential for substantial societal impact.

Similarly, for business leaders, the rise of virtual tutors could signal new market opportunities. Companies that can develop effective, ethical, and scalable tutoring solutions will find themselves at the forefront of the education sector. This could lead to partnerships with educational institutions, governments, and non-profits aiming to make quality education accessible to all.

The evolution of virtual tutors also aligns with the broader trends in AI and machine learning, as discussed in previous chapters. The advancements in these fields are not isolated; they are part of a complex, interconnected web of technological progress that spans across various industries. By understanding how these technologies interplay, one can appreciate the holistic impact they will have, not just on education but on society as a whole.

In essence, virtual tutors represent a paradigm shift in how we approach learning. They offer a solution to some of the most pressing educational challenges of our time. Yet, they also come with their own set of questions and concerns that need careful consideration. As we navigate this new landscape, it is crucial to balance innovation with ethical responsibility, ensuring that the benefits of these technologies are accessible to all.

Ultimately, the future of AI in education, exemplified by virtual tutors, is a testament to human ingenuity and the relentless pursuit of progress. It is a journey filled with both promise and pitfalls, but one that holds the potential to transform lives across the globe. As we continue to explore and refine these technologies, the goal remains clear: to create a more inclusive, efficient, and enriching educational experience for everyone.

Chapter 13:
AI in Environmental Sustainability

As we grapple with the accelerating urgency of climate change, AI emerges as a beacon of hope, illuminating pathways to a more sustainable future. Imagine a world where advanced climate modeling anticipates devastating weather patterns well before they wreak havoc, allowing communities to brace and adapt. Picture AI-driven resource management systems optimizing water usage in drought-stricken regions and intelligently regulating energy consumption in bustling urban centers. Not limited to mere data crunching, AI has the transformative potential to decode nature's complex signals and guide us in restoring Earth's delicate balance. This isn't just a vision of what could be—real-world applications are already beginning to show remarkable successes. In harnessing AI's power, we are not only innovating but also stewarding the planet for generations to come. It's about creating symbiosis between cutting-edge technology and timeless environmental wisdom, paving the way for a resilient and sustainable Earth.

Climate Modeling

When you think of climate change, vast, complex interactions within Earth's systems probably come to mind. Maybe it's images of melting glaciers, rising sea levels, or extreme weather events. What if I told you that artificial intelligence (AI) is now diving into this intricate web to help us understand, predict, and perhaps mitigate the impacts of climate change?

Climate modeling involves simulating Earth's climate system using mathematical representations. These models have been around for decades, offering insights into how variables like temperature, precipitation, and atmospheric pressure interact. However, traditional climate models are often limited by their ability to process the staggering amount of data required for accurate predictions. That's where AI steps in, and the transformative potential is enormous.

AI algorithms, often leveraging machine learning, offer an unprecedented ability to analyze vast amounts of climate data quickly and accurately. By doing so, they can identify patterns and trends that might elude traditional methods. The result? More accurate, nuanced climate models that can help scientists make better predictions about future climate scenarios.

Think about the ramifications of more precise climate models. For businesses, it means better risk assessment and planning. A company can anticipate how climate change might impact its supply chain, operations, and market conditions. For governments, improved models mean more effective policies and responses to natural disasters. Early warnings can save lives and resources.

Let's not forget the motivational spark behind this work. The insights gained from AI-driven climate modeling could act as powerful catalysts for change. They provide tangible evidence of climate trends and potential future impacts, moving the issue from abstract projections to concrete realities. This kind of evidence can push policymakers, corporations, and even individuals to take decisive action.

Consider the role machine learning plays in refining these models. Historically, climate simulations required supercomputers crunching equations over days, weeks, or even months. Machine learning algorithms can speed this process up by identifying shortcuts and optimization strategies. They continually learn from new data, improving their accuracy over time—a self-improving loop that's invaluable in climate science.

One remarkable application involves using AI to downscale global climate models to regional or even local levels. These downscaled models help communities understand specific risks and plan accordingly. For instance, coastal towns can get detailed forecasts on rising sea levels and storm surges, letting them bolster defenses and prepare evacuation plans.

But AI's capabilities go beyond just faster calculations. The sophisticated models can integrate disparate types of data—from satellite imagery and oceanic temperatures to humidity levels and deforestation rates. The more diverse the data, the richer the insights, creating a multifaceted view of our climate reality. This integration of various data sources lets scientists build models that are not only more comprehensive but also more resilient to uncertainties.

Moreover, AI empowers the climatic research community to run multiple simulations with different variables, providing a broader range of potential outcomes. This range helps in understanding the probabilistic impacts of climate patterns, enabling a deeper and more nuanced comprehension of our planet's future.

Take, for example, the predictive power of AI in understanding extreme weather events. Traditional models often struggle with the erratic nature of phenomena like hurricanes or heatwaves. AI can dissect these complexities more effectively, offering better forecasts and preparation measures. This preparedness can significantly mitigate the damage and save lives, giving humanity a fighting chance against nature's fury.

There's also a profound educational aspect to this technology. AI-driven climate models can serve as an educational tool to present climate data in engaging, understandable ways. Interactive visualizations and simulations offer new avenues for educating the public about climate issues, making the science behind climate change more accessible and persuasive.

The technological progress we're seeing isn't just theoretical; it has real-world applications that are already making a difference. In agriculture, AI-driven models help farmers anticipate and adapt to weather patterns, optimizing crop yields and minimizing waste. This isn't just about resource management; it's about feeding the world in the face of increasingly volatile climate conditions.

In the energy sector, renewable energy forecasting is becoming more precise. Solar and wind energy outputs, for example, are dependent on weather conditions. AI can improve the prediction accuracy of these variables, allowing for better grid management and more efficient use of resources.

One can't ignore the collaborative nature of these advancements. Scientists, technologists, and policy-makers are coming together like never before to share data, methodologies, and discoveries. This interdisciplinary collaboration fosters innovation and accelerates the pace at which solutions are found and applied.

However, applying AI to climate modeling isn't without its challenges. Ethical considerations around data privacy, representation, and algorithmic bias need to be addressed. For AI to be a trusted tool in climate science, transparency in data usage and algorithms is critical. Stakeholders must ensure that these models are inclusive and fairly represent different regions and communities, particularly those most vulnerable to climate impacts.

Despite the challenges, the impact of AI on climate modeling offers incredible promise. It's a robust tool that not only aids in understanding our current climatic conditions but also equips us with the foresight needed to shape a sustainable future.

So, why should tech enthusiasts and business leaders care? Because AI in climate modeling is more than just advanced technology at play. It's a crucial early warning system, a risk management tool, and perhaps most importantly, a wake-up call that there's still time to make a difference. With AI, we're not just observing the planet's slow spiral into climate chaos; we're actively working to turn things around.

In a world where the repercussions of climate change are becoming increasingly unavoidable, leveraging AI to enhance climate modeling is an inspiring testament to humankind's resilience and ingenuity. It's an exciting frontier where technology and environmental consciousness meet, providing a beacon of hope and a path toward a more sustainable future.

As we continue to innovate and integrate AI into climate modeling, the potential to avert climate disaster becomes ever more tangible. We're at the cusp of being able to use machine intelligence to safeguard our planet's future. Now, that's a vision worth pursuing.

Resource Management

Think of AI as more than just a tool; see it as an ally in the battle against resource mismanagement. With Earth's resources dwindling and environmental concerns escalating, leveraging AI becomes vital for more judicious use of what we have left.

From monitoring deforestation to predicting water shortages, the role of AI in resource management is expanding in ways that deeply impact our daily lives and future economies. For instance, AI algorithms can sift through complex data sets to detect patterns that would take humans years to uncover. This isn't just about saving time—it's about finding solutions before problems escalate beyond control.

Consider water management. Smart sensors coupled with AI can monitor and predict how water is used in urban areas versus agricultural sectors. This can lead to better allocation and reduced waste. Instead of waiting for a drought to hit, predictive analytics can help communities prepare in advance, ensuring that water is directed where it's needed most. By avoiding scarcity, we also prevent the social and economic chaos that often accompanies it.

One fascinating application is in precision agriculture. Through AI-driven insights into soil health and crop performance, farmers can make real-time adjustments to planting, watering, and harvesting. This kind of tailored approach not only maximizes yield but also minimizes input costs and environmental impacts. Imagine a future where every drop of water and every grain of fertilizer is used as efficiently as possible. That's the promise AI holds.

Forests, often called the lungs of our planet, also benefit from AI surveillance. Satellite imagery analyzed by machine learning models can track illegal logging activities. Governments and organizations can then respond swiftly to protect these vital resources. The technology goes beyond mere detection; AI can predict where illegal activities might happen next, enabling a proactive stance rather than a reactive one.

When it comes to renewable energy, AI's role becomes incredibly intricate and impactful. Wind turbines, solar panels, and hydroelectric plants generate massive amounts of data. AI models can analyze this data to predict maintenance needs, optimize energy storage, and even forecast weather conditions to maximize efficiency. For example, solar farms can utilize AI to predict cloud cover, adjusting operations to store energy when production is high and release it when production dips.

Supply chain management also gains a transformative edge through AI. With its ability to analyze vast amounts of data rapidly, AI can identify inefficiencies and recommend improvements. Whether it's optimizing routes for delivery trucks to save fuel or managing inventory to reduce waste, smarter supply chains mean lower carbon footprints and cost savings for businesses.

It's easy to overlook the impact of AI on something as mundane as waste management, but here, too, AI is making strides. Smart bins equipped with sensors can sort recyclables from waste, reducing human error and increasing recycling rates. AI systems can also analyze consumption patterns to predict waste volumes, allowing for better planning and more efficient waste collection routes.

AI's role in managing natural disasters can't be underestimated. Advanced algorithms can analyze seismic data to predict earthquakes or monitor ocean temperatures to forecast hurricanes. Early warning systems equipped with AI can save lives and help allocate resources more effectively during emergencies. These predictive analytics not only prepare us for immediate threats but also help in long-term planning for sustainable living.

One often-mentioned but critically important application is air quality management. Cities grappling with pollution can use AI to monitor air quality in real time, predict pollution spikes, and suggest measures to mitigate them. Whether through smart traffic management systems designed to reduce emissions or AI models predicting industrial contributions to air pollution, the end goal remains the same: cleaner air for healthier communities.

Energy consumption in buildings is another frontier. AI can adjust lighting, heating, and cooling in real time based on occupancy and weather conditions. These "smart buildings" use energy more efficiently, reducing both costs and environmental impacts. Imagine a world where skyscrapers aren't just monuments of steel and glass but active participants in energy management, making decisions that benefit both the environment and the budget.

Now, picture integrated ecosystems where various AI applications interact seamlessly. Water management systems could connect with agriculture AI, waste management could tie into smart city frameworks, and renewable energy AI could link with national grids. This interconnected web of AI-driven systems could create a harmonious, efficient, and sustainable society capable of managing resources in ways that previously seemed unattainable.

Resource management isn't just about using AI; it's about integrating it thoughtfully into existing systems while preparing for future complexities. The ultimate goal is to create a balanced ecosystem where human ingenuity and machine efficiency coexist, driving us toward a more sustainable future. What if we could make resource abundance, not scarcity, the norm? The key lies in leveraging AI as a crucial piece of this grand puzzle.

As businesses, governments, and individuals embrace this future, it's critical to have strategies for rolling out these AI systems. Training, ethics, and data management can't be afterthoughts. We need frameworks to ensure these tools are used responsibly and effectively. And though collaborative efforts across sectors and nations will be required, the effort promises untold rewards in efficiency, sustainability, and equity.

In conclusion, AI's transformative potential in resource management is both a technological marvel and a moral imperative. By integrating these advanced systems into our daily lives and industries, we get one step closer to a world where resources are abundant, conflicts over scarcity are reduced, and the environment thrives. The promise of AI in resource management isn't just about technological innovation; it's about building a better, more sustainable world for future generations.

Chapter 14:
AI and Security

As we navigate deeper into the AI era, security emerges not just as a priority but as the very bedrock of trust and progress. The interplay between AI and security is intricate, blending promise and peril. On one hand, artificial intelligence supercharges our defenses with predictive analytics that can spot cyber threats before they unfold, and on the other, it arms bad actors with unprecedented tools for sophisticated attacks. Imagine a future where AI algorithms safeguard national defense systems and banks, but also where rogue AIs could orchestrate breaches that bypass traditional cyber barriers. It's a double-edged sword that demands innovation and vigilance. The balancing act here isn't merely technical—it's deeply human, requiring ethical considerations and a new breed of tech warriors who can anticipate and outwit the evolving landscape of threats. As you venture into this chapter, get ready to unravel both the fortresses and the vulnerabilities that define AI-powered security. It's a dynamic realm where each advancement in technology brings new strategies, and every challenge becomes a catalyst for breakthrough solutions.

Cybersecurity

As AI becomes more integrated into our daily lives, its implications for cybersecurity become critical. Modern cybersecurity faces a rapidly evolving threat landscape, one where attackers and defenders are locked in a constant game of cat and mouse. AI can change the dynamics of this game dramatically.

Let's start with the basics. Cybersecurity refers to the practice of protecting systems, networks, and programs from digital attacks. These cyberattacks are often aimed at accessing, changing, or destroying sensitive information; extorting money; or disrupting normal business processes. AI, with its ability to analyze enormous datasets, predict patterns, and adapt quickly, embodies a promising tool for enhancing these protective measures.

Firstly, AI brings a level of speed and accuracy to threat detection that was previously unavailable. Traditional security systems rely on predefined rules and signatures to spot threats, which means they often lag behind new and sophisticated attacks. But AI, especially machine learning algorithms, can identify anomalies and potential threats in real-time, minimizing the window for malicious activities by recognizing patterns that deviate from the norm.

Consider how machine learning could revolutionize phishing detection. Phishing attacks, which trick users into providing personal information through fake emails or websites, are increasingly sophisticated. An AI system can be trained on thousands of examples of phishing communications and then deployed to scan incoming emails, flagging potentially dangerous ones with impressive accuracy.

The evolution of cybersecurity isn't just about better detection, though. Automated response systems are another arena where AI shines. Once a threat is detected, quick containment is crucial. AI can automate these responses, ensuring that actions like shutting down affected systems, isolating malware, and patching vulnerabilities occur instantly, reducing potential damage significantly.

However, it's not all sunshine and roses. The emergence of AI in cybersecurity introduces new complexities. Cybercriminals are also leveraging AI to enhance their attacks, creating more sophisticated malware and phishing campaigns that are harder to detect. This raises the stakes in a digital arms race, where both sides are constantly evolving and adapting AI technologies.

One poignant example of AI being used maliciously is in the development of AI-powered malware. These advanced forms of malware can learn, adapt, and avoid detection based on the cybersecurity measures they encounter. Such malware can morph its patterns upon detection to regain stealth, making it a moving target that traditional cybersecurity measures struggle to keep up with.

This dual use of AI, by both attackers and defenders, means cybersecurity strategies must continuously evolve. It calls for a robust and dynamic approach to fortify defenses. Business leaders, tech enthusiasts, and policy makers need to collaborate, ensuring they stay updated with the latest advancements and threats in the AI-cyber-security landscape.

To navigate this evolving threat landscape effectively, companies and individuals must leverage AI to build more resilient cyber defenses. It's not just about having the latest tools but also about cultivating a mindset that embraces continuous learning and adaptation. Regular updates, urgent patches, and comprehensive cybersecurity training are indispensable components of a strong defense strategy.

Moreover, AI can significantly aid in vulnerability management. Vulnerability management involves identifying, evaluating, treating, and reporting on security vulnerabilities in systems and the software that runs on them. AI can automate these processes, conducting continuous assessments to identify latent vulnerabilities and prioritizing them based on the level of threat they pose. This allows security teams to focus their efforts on the most critical issues first.

Additionally, AI can enhance the concept of predictive maintenance in cybersecurity. Predictive maintenance usually applies to physical machinery, forecasting failures before they occur. When adapted to cybersecurity, AI can anticipate potential security breaches by analyzing patterns and trends over time. This proactive approach can preemptively address issues, thereby fortifying defenses well before any real damage is done.

Discussing cybersecurity also brings us to the importance of encryption and how AI can optimize it. Encryption is the process of encoding information so that only authorized parties can access it. As quantum computing looms on the horizon, current encryption methods may become obsolete. AI can assist in developing new encryption algorithms that are more resistant to quantum attacks, ensuring data privacy and integrity remain intact in the future.

The role of AI in cybersecurity is not limited to protecting digital assets alone; it extends to safeguarding physical systems too. With the advent of the Internet of Things (IoT), the line between digital and physical security is increasingly blurred. Connected devices—from smart refrigerators to industrial control systems—pose new vulnerabilities. AI can monitor these IoT networks, identifying and mitigating threats that target both physical and digital realms.

In the multifaceted world of cybersecurity, collaboration is paramount. AI tools, no matter how advanced, need to be part of a broader, integrated security strategy. This strategy should include human expertise, industry standards, and collaborative efforts between public and private sectors. AI can identify threats and suggest countermeasures, but human intuition and decision-making processes are critical to interpret and act on these insights effectively.

Regulation and policy also play a pivotal role. Governments and regulatory bodies must craft policies that encourage the ethical and responsible use of AI in cybersecurity. This includes ensuring that the

AI tools themselves are secure, their development transparent, and their usage abiding by privacy laws and ethical considerations.

Public and private sectors will need to work hand in hand, exchanging threat intelligence and developing unified strategies to mitigate risks. Global cooperation could help standardize best practices and frameworks, enabling a more cohesive response to the global threat landscape.

The journey towards robust AI-enhanced cybersecurity is challenging but brimming with possibilities. As we focus on defending our digital borders with AI, we must remain vigilant, adaptive, and collaborative. We stand on the brink of a new era where AI is not just a tool but a partner—empowering us to build a safer, more secure digital world.

National Defense

In the realm of national defense, artificial intelligence (AI) is not just a promising tool; it's a transformative force. Historically, defense strategies have always adapted to incorporate the latest technological advancements, and AI is no exception. Today's military and defense agencies are looking to AI for its potential to revolutionize everything from decision-making to battlefield tactics and cybersecurity.

One of the most exciting applications of AI in national defense is its ability to process vast amounts of data quickly and accurately. This capability is particularly crucial in intelligence gathering, where time-sensitive information can mean the difference between success and failure. Machine learning algorithms can sift through terabytes of data from various sources—satellites, drones, social media—identifying patterns and anomalies that would be almost impossible for human analysts to detect promptly.

Imagine a scenario where AI algorithms analyze satellite images to monitor troop movements in real-time, providing actionable intelligence almost instantaneously. These systems can identify subtle changes in terrain, detect new equipment deployments, or even anticipate future moves using predictive analytics. They act as force multipliers, significantly enhancing the capabilities of human analysts and tactical decision-makers.

No discussion on AI and national defense would be complete without touching on autonomous systems, particularly unmanned aerial vehicles (UAVs) and autonomous ground vehicles. These systems can operate in environments that are too dangerous for human soldiers, reducing casualties and increasing the efficiency of operations. Advanced AI allows these autonomous systems to make split-second decisions, tackle complex tasks, and adapt to changing conditions on the battlefield.

The ethical implications of deploying lethal autonomous weapons, or "killer robots," can't be ignored. There's an ongoing debate about the level of autonomy these systems should possess, with concerns about accountability and the potential for unintended consequences. While technology races ahead, international laws and ethical frameworks lag behind, creating a precarious situation that demands attention from policy-makers and military leaders alike.

Another vital area where AI contributes significantly is in cybersecurity. Defense systems are increasingly reliant on intricate networks and sophisticated software, making them prime targets for cyber attacks. AI can play a dual role here: both as a defender and as an attacker. On the defensive side, machine learning algorithms can monitor network activity for suspicious behavior, flagging potential threats in real-time. These systems can learn from each incident, becoming more adept at identifying and thwarting attacks over time.

Conversely, AI can also be used offensively in cyber warfare. AI-designed programs can probe enemy networks for vulnerabilities, initiate attacks that adapt and change based on the target's defenses, and even conduct psychological operations through misinformation campaigns. The dynamic nature of AI could make these cyber tools exceptionally difficult to counter, posing challenges for traditional defense mechanisms.

It's essential to recognize the human element in integrating AI into national defense. While AI can enhance and augment human capabilities, it won't replace the need for skilled personnel who understand the nuances of warfare and strategic planning. Training programs must evolve to prepare military personnel to work alongside AI systems effectively. This involves cross-disciplinary knowledge, combining traditional military training with a deep understanding of data science and machine learning.

AI's impact on defense logistics and supply chain management shouldn't be overlooked. Advanced algorithms can optimize supply routes, predict maintenance needs for military equipment, and manage inventory more efficiently. By ensuring that resources are allocated where they are most needed, AI can significantly enhance operational readiness and reduce waste.

The geopolitical landscape will also be shaped by AI advancements in national defense. Countries that effectively harness AI technologies will likely gain a strategic advantage, redefining global power dynamics. This situation leads to an AI arms race, where nations vie to develop the most advanced capabilities, raising questions about global stability and the potential for conflict.

International collaboration and treaties will be crucial in managing the implications of AI in national defense. Establishing norms and agreements on the use of AI in warfare can help mitigate risks and promote stability. It's a complex issue involving diplomacy, trust-building, and a shared commitment to ethical standards.

Corporate involvement in AI development for national defense is another significant trend. Tech companies specializing in AI research are increasingly partnering with defense agencies, contributing their expertise to advance military applications. These partnerships can accelerate innovation but also raise ethical and transparency concerns. Balancing corporate interests with national security priorities will be an ongoing challenge.

Looking to the future, AI's role in national defense is poised to expand even further. We're likely to see advancements in AI-driven strategic simulations, where algorithms can model various conflict scenarios and predict their outcomes, offering valuable insights for policy and decision-makers. Additionally, AI-assisted command and control systems could provide real-time data fusion from multiple sources, enhancing situational awareness and decision-making in complex environments.

While AI offers tremendous potential in national defense, it also presents significant risks. Unintended consequences, such as algorithmic biases or system failures, could have catastrophic outcomes when applied to military operations. Rigorous testing, validation, and ethics oversight will be essential to ensure AI systems perform as expected and do not exacerbate existing threats.

In conclusion, AI is reshaping the landscape of national defense in profound ways. Its potential to enhance intelligence gathering, optimize logistics, and revolutionize battlefield tactics is matched by ethical and strategic challenges that require careful navigation. As we move forward, combining technological innovation with ethical considerations and international cooperation will be crucial to harnessing AI's benefits while mitigating its risks. The future of national defense depends not just on technological prowess but also on our ability to use these tools responsibly and wisely.

Chapter 15:
Societal Impacts of AI

As we dive into the societal impacts of AI, it's clear we're standing on the precipice of a revolution that's rewriting the rules of the game. AI isn't just a technological marvel; it's a cultural shift poised to reshape everything from social structures to global perceptions. On one hand, there's the promise of unparalleled efficiency and capability, heralding an era where repetitive tasks are automated and creativity is unleashed. On the other, the specter of social inequalities looms large, fueled by questions about accessibility and the concentration of technological power. The public's perception of AI oscillates between fascination and fear, driven by media portrayals and personal experiences. This duality—optimism tempered by caution—defines our current relationship with AI, urging us to navigate this transformative landscape with both enthusiasm and prudence. As AI continues to permeate every facet of society, understanding its broader impacts isn't just advisable; it's essential for crafting a future that benefits all.

Social Inequality

We've journeyed through the landscape of AI's vast capabilities and its applications in various sectors. Now, let's zero in on one of the more daunting issues arising from this technological evolution—social inequality. The way AI is being integrated into society has the potential to both bridge gaps and widen them. This duality is deeply rooted in how access to AI technologies and their benefits are distributed.

Let's start with education. AI-driven tools like adaptive learning systems and virtual tutors have the potential to transform education. They can provide personalized learning experiences tailored to each student's needs, boosting educational outcomes. However, there's a catch: these advanced tools are not equally available to everyone. Schools in affluent areas, with ample funding, can afford to implement cutting-edge AI technologies, while underfunded schools struggle to provide basic technological infrastructure. This disparity only exacerbates existing educational inequities.

When it comes to the workforce, the scenario is no less complex. AI fuels automation, and with it comes job displacement. Certain jobs, especially those involving repetitive tasks, are more susceptible to being automated. Workers in these positions often come from lower-income backgrounds, with limited opportunities to retrain or pivot to new career paths. While new opportunities are undoubtedly created, they frequently require advanced skills and education. The net result can be an economic divide where the rich get richer, benefiting from AI, and the poor get poorer, displaced by it.

Consider healthcare—a sector where AI has made remarkable strides. AI diagnostic tools and personalized medicine promise better health outcomes. Nevertheless, these advancements are often expensive and inaccessible to underserved communities, amplifying health disparities. It's a bitter irony that those who could benefit the most from AI-driven healthcare solutions are often those who can't afford them.

Access to AI is not merely a matter of economic disparity; it's also about digital literacy. As AI permeates our daily lives, the ability to understand and engage with these technologies becomes crucial. However, digital literacy is unevenly distributed, creating a digital divide. Those with limited access to technology and digital education are left out of opportunities that could uplift them, be it in education, employment, or healthcare.

Let's talk about bias—an often overlooked yet critical issue. AI systems, many of which learn from historical data, can inadvertently inherit and perpetuate societal biases. For example, recruitment algorithms may unfairly disadvantage certain demographic groups based on historical hiring practices. Similarly, facial recognition technologies have shown higher error rates for people of color. When biased AI systems are deployed in decision-making processes that affect lives—like hiring, lending, or law enforcement—the stakes are incredibly high. Correcting these biases involves more than just technical tweaks; it calls for a fundamental reevaluation of how we gather and interpret the data that fuels AI.

One might ask, "Is there a way to mitigate these inequalities?" The answer isn't straightforward. Policies aimed at equitable AI deployment need to be comprehensive and multifaceted. They should include measures to ensure universal digital literacy, equitable access to advanced technologies, and the implementation of strict guidelines for fairness and transparency in AI systems. More importantly, these policies should be crafted with input from a diverse array of stakeholders, ensuring that voices from marginalized communities are heard and considered.

Another promising avenue is public-private partnerships. Governments, educational institutions, and private companies can collaborate to provide broader access to AI technologies and education. For instance, tech companies can invest in community programs that teach AI-related skills to underprivileged populations. Governments can implement subsidies or incentive programs to make AI tools more affordable for schools and healthcare providers in low-income areas.

Let's not forget the role of grassroots movements and local initiatives. Community-driven efforts to bridge the digital divide can make a substantial difference. Programs that offer coding classes, tech workshops, and access to digital resources empower individuals, equipping them with skills to participate in the AI-driven economy. These initiatives also build community resilience, fostering a supportive environment where people can learn and grow together.

We also need to consider the role of legislation in shaping an equitable AI landscape. Regulation should not stifle innovation but rather guide it towards inclusivity. Laws focused on preventing algorithmic bias, ensuring data privacy, and promoting transparency are crucial. Additionally, governments must hold companies accountable, requiring them to demonstrate how their AI systems align with principles of fairness and equity.

Of course, addressing social inequality is not a one-time effort but an ongoing commitment. As AI technologies evolve, so too must our strategies to ensure they uplift rather than divide. Continuous research into the societal impacts of AI is essential, as is the public's engagement in discussions about how these technologies should be used. We're all stakeholders in this AI-driven future, and our voices can shape a more just and equitable world.

So, where do we go from here? The challenges are substantial, but they are not insurmountable. With a concerted effort from all sectors of society, we can harness the potential of AI to create opportunities rather than exacerbate divisions. By prioritizing equity and inclusion in our approach to AI, we can navigate this transformative period with a vision of a future where technology serves all of humanity, not just a privileged few.

As we continue to delve into the societal impacts of AI, it's crucial to remember that technology, in itself, is neutral. It's how we implement and regulate it that will determine whether it becomes a tool for uplifting or dividing. Let's take this moment to reflect on the kind of future we want to create, leveraging AI not just for advancement, but for greater equity and social justice.

Public Perception

The way the public perceives AI is a fascinating blend of curiosity, skepticism, excitement, and fear. For some, AI represents the dawn of a utopian future where machines handle mundane tasks, leaving humans free to pursue creativity and leisure. Others foresee a dystopian nightmare where robots run amok, human jobs vanish, and privacy becomes a quaint relic of the past. This duality in perception often hinges not just on the capabilities of AI, but on how these capabilities are communicated, demonstrated, and understood.

To put it simply, public perception of AI is deeply influenced by the media. Movies, books, and news outlets frequently shape how AI is viewed by millions. Remember the menacing HAL 9000 from "2001: A Space Odyssey"? Or the benevolent C-3PO from "Star Wars"? These iconic representations have left indelible marks on how people imagine AI: as either a villain or a hero. The truth, however, is much more nuanced.

Survey results suggest that many people remain unclear about what AI really is. Is it a super-intelligent, sentient being? Or just a set of complex algorithms and data processes? This murkiness often leads to misconceptions. A study at Stanford University found that a significant portion of the public believes AI can think and make decisions independently. While advanced algorithms can indeed perform intricate tasks, they're far from having the self-awareness or moral reasoning that humans possess.

It's also essential to consider the context in which people encounter AI. For instance, AI in healthcare—like diagnostic tools and personalized medicine—is generally met with enthusiasm and hope.

People appreciate the potential for earlier disease detection and tailored treatments. Conversely, the use of AI in surveillance and data monitoring often triggers anxiety and concerns over privacy and civil liberties. These multifaceted reactions reveal that public perception isn't monolithic; it's context-driven and highly variable.

Another crucial factor is education—or the lack thereof. A more informed public tends to have a balanced view of AI. They recognize both its benefits and limitations. Despite the availability of information, a knowledge gap persists, making it easy for fear and misinformation to spread. Educational initiatives begin to bridge this gap, slowly altering perceptions positively. Workshops, online courses, and public seminars are excellent strategies to demystify AI.

Now, let's talk about the impact of personal experiences with AI. Who hasn't been greeted by a chatbot on a customer service website or received a product recommendation from an e-commerce giant? These daily encounters with AI-powered systems often shape opinions more profoundly than abstract theories or news articles. When these interactions are smooth and helpful, public sentiment skews positively. But a frustrating or invasive encounter can reinforce negative stereotypes, casting AI in a poor light.

The tech industry plays a significant role in molding public perception too. Companies often market AI as the next big thing, highlighting breakthroughs like self-driving cars or advanced language processors. However, when these ambitions fall short—take the numerous high-profile failures of autonomous vehicle trials, for instance—the gap between expectation and reality becomes glaringly evident, leading to disillusionment.

Public perception also intersects with socio-economic factors. For someone in a high-paying tech job, AI might symbolize progress and wealth. However, for someone in a low-wage job fearing automation, AI could appear as a direct threat to livelihood. This divergence implies that public sentiment isn't just a matter of informational clarity but also emotional and economic context.

It's worth mentioning the ethical quandaries that surface in public discussions of AI. Concerns about bias in AI algorithms, data privacy, and moral accountability spark fierce debates. People often worry whether AI will be used responsibly or if it will perpetuate existing inequalities. These concerns aren't unfounded, as evidenced by instances where AI systems have demonstrated bias in hiring practices or sentencing guidelines. While the tech community works avidly to address these issues, public trust remains a fragile commodity.

Moving forward, transparency is key. When organizations are open about how their AI systems work, what data they use, and how they make decisions, they foster trust. This transparency also involves acknowledging and addressing failures. Instead of covering up mistakes, owning them and demonstrating efforts to improve can significantly tilt public perception in a more trustworthy direction.

Regulations and policies further shape public views on AI. Clear rules provide a sense of security, assuring people that there are checks and balances in place. When governments and regulatory bodies step in with thoughtful legislation, it adds a layer of accountability that resonates positively with the public. Without stringent regulations, perceptions can skew towards fear and distrust.

Community involvement can also play an influential role. People are more likely to view AI positively when they're involved in its development and deployment. Initiatives that enable public participation, such as town hall meetings and community advisory boards, help demystify AI. They allow people to voice their concerns and contribute to the conversation.

Finally, let's not underestimate the power of positive storytelling. Highlighting success stories where AI has made a real difference—

improving medical diagnoses, reducing traffic accidents, or helping combat climate change—can be incredibly motivating. These stories instill hope and inspire a sense of collective progress. When people see AI improving lives, it shifts from an abstract concept to a tangible tool for good.

As we step deeper into an AI-driven future, it's clear that public perception will continue to evolve. This evolution won't be linear. It will oscillate between extremes, driven by new developments, regulatory changes, and societal needs. But by fostering transparency, encouraging education, and involving communities, we can shape a balanced and well-informed public outlook on AI.

So, the next time you encounter AI, whether it's a friendly recommendation engine or an autonomous car, remember that your perception is part of a broader, ever-changing narrative. It's a narrative we all contribute to, consciously or not, shaping the role AI will play in our collective future.

Chapter 16:
AI's Role in Innovation

Innovation has always been the lifeblood of progress, but with AI, we're unleashing creativity and problem-solving on an epic scale. Imagine a world where research and development cycles, traditionally measured in years, are now accelerated multifold by AI-driven insights and predictions. This isn't just a futuristic dream—it's happening now. AI's capability to sift through colossal datasets with precision is transforming patent analysis, enabling businesses to pinpoint trends and emerging technologies with laser accuracy. Whether it's discovering new materials, fine-tuning processes, or predicting market needs, AI is at the forefront, turning ambitious ideas into tangible realities faster than ever. The synergy between human ingenuity and AI's computational power is not just propelling us forward; it's rewriting the very rules of innovation. It's like adding jet fuel to the engine of progress, pushing boundaries and dissolving the limitations of what we thought was possible.

Research and Development

Research and Development (R&D) represents the foundational element where the transformative power of AI truly shines. It's in the painstaking process of experimentation, discovery, and iteration that AI's potential becomes undeniable. Whether in massive universities, agile startups, or corporate R&D departments, AI is a catalyst driving innovation and propelling us into uncharted territories. But how exactly is AI making waves in this space?

Let's start simple. Think about the scientific method. Every breakthrough begins with a hypothesis, followed by systematic experimentation. Traditionally, this would involve data collection, analysis, and interpretation – a slow, meticulous process. Enter AI. With machine learning (ML) algorithms, researchers now automate data analysis, crunching vast datasets in a fraction of the time humans can. This means hypotheses can be tested, refined, and validated much faster.

Consider the recent advances in drug discovery. By leveraging AI, pharmaceutical companies can predict how different compounds will interact with biological targets. It's like having a digital oracle that scans molecular structures and predicts their efficacy. This doesn't just speed up the research cycle; it also cuts costs significantly. Before AI, it might have taken a decade and billions of dollars to bring a drug to market. Now, that timeline can be slashed dramatically, bringing life-saving treatments to patients faster.

In the realm of materials science, AI's impact is equally groundbreaking. Researchers are using AI to discover new materials with specific properties, leading to advancements in everything from battery technology to aerospace engineering. AI-driven simulations can predict how a material will behave under certain conditions, allowing scientists to tweak properties on the fly. It's a rapid feedback loop that accelerates material innovation like never before.

Beyond these specific fields, AI is also being applied to enhance R&D processes universally. Natural language processing (NLP) algorithms sift through vast swathes of scientific literature, journals, and patents to extract relevant information. This helps researchers stay on top of the latest developments and ensures that their work builds on previous discoveries rather than duplicating them. Imagine having a research assistant who has read every scientific paper ever written – that's the power AI brings to the table.

Collaborative AI platforms are another burgeoning area. These are systems designed to facilitate cooperation between different research teams, often spread across continents. By sharing data, models, and findings in real-time, researchers can achieve breakthroughs faster. The result is a more connected, informed, and efficient R&D ecosystem where innovation isn't confined by geographical borders.

Of course, this influx of AI-driven R&D raises interesting ethical and practical questions. Who owns the intellectual property when an AI contributes to a discovery? Is it the organization that trained the AI, the developers who built the algorithms, or the researchers using the AI? Moreover, there are concerns about transparency and bias. AI models are only as good as the data fed into them. If that data carries inherent biases, the results can reinforce existing inequities in scientific research.

Then there's the matter of accessibility. Advanced AI tools often require substantial computational resources and expertise, potentially widening the gap between well-funded labs and smaller research entities. Democratizing access to these tools remains a challenge, but it's one that the tech community is keenly aware of. Open-source AI frameworks and collaborations between academia and industry are steps in the right direction.

So, why is all of this important? Why should tech enthusiasts, business leaders, and futurists care about AI's role in R&D? Simply put, AI doesn't just accelerate innovation; it expands the realm of what's possible. By pushing the boundaries of current knowledge and capability, AI facilitates a future where today's sci-fi dreams become tomorrow's reality. Think of clean energy solutions, personalized medicine, resilient infrastructure, and beyond – these are all within reach, thanks to AI-powered R&D.

Moreover, the iterative nature of AI models means they only get better with time. As more data is fed into these systems, they become increasingly adept at discerning patterns and making predictions. So not only are we innovating faster, but we're also improving the very tools that drive this innovation. This creates a virtuous cycle where success begets success, leading to exponential growth in knowledge and technological capabilities.

Let's not forget the critical role of multidisciplinary approaches. AI thrives on data, but the insights gleaned from that data often come from combining expertise from various fields. Whether it's bioinformatics, computational physics, or social sciences, integrating AI into R&D necessitates breaking down silos and fostering cross-disciplinary collaboration. This convergence is the bedrock of truly groundbreaking discoveries.

Ultimately, the adoption of AI in R&D isn't just a technological shift; it's a paradigm shift. It demands a new way of thinking about problems and solutions. Flexibility, curiosity, and openness to new methods and ideas become paramount. While the journey involves challenges – ethical concerns, resource allocation, and skill gaps – the destination promises a landscape rich in innovation and progress.

In conclusion, as AI continues to embed itself deeply in the R&D landscape, it holds the promise of accelerating the cycle of knowledge creation and technological advancement. For anyone involved in tech, business, or future-oriented thinking, understanding AI's role in R&D is crucial. It offers a lens through which we can envision and navigate the rapid changes on the horizon. The future is not just something that happens to us; it's something we actively shape – and with AI, we have one of the most powerful tools to do so.

Patent Analysis

Let's dive into the fascinating world of patents and how AI is revolutionizing the innovation landscape. Patents are more than just legal documents; they represent the pinnacle of human creativity. They encapsulate breakthroughs that push the boundaries of what's possible. In today's fast-paced environment, AI has become a catalyst in both the formation and accumulation of patents, sparking a new wave of technological advancements.

It's intriguing to observe how AI's impact isn't limited to mere invention but extends to the entire lifecycle of a patent. From ideation to filing and even enforcement, AI tools are integrated at each step to make processes faster and more efficient. Imagine an AI system that can swiftly analyze a vast amount of prior art, enabling inventors to spot gaps or opportunities in the existing reservoir of knowledge. This capability not only accelerates the process but substantially enhances creativity and originality.

AI's prowess in data analytics has forever changed the patent landscape. By leveraging machine learning algorithms, companies can analyze trends and predict future technology trajectories. This foresight can be incredibly valuable for businesses deciding where to focus their R&D efforts. Essentially, AI is turning patent analysis from a reactive practice into a proactive strategy. The predictive power of AI can help companies identify potential white spaces—those untouched territories ripe for innovation and patenting.

Patent filing is another area where AI is making waves. Traditionally, this process has been laborious and time-consuming, involving extensive documentation and legal scrutiny. AI-driven tools now assist in drafting patent applications, identifying the right keywords, and ensuring that the language complies with legal norms. This reduces human error and speeds up the filing process. These AI tools can even compare applications against millions of other documents to ensure novelty and minimize infringement risks.

The review process conducted by patent offices is also undergoing a transformation courtesy of AI. Patent examiners are empowered with AI-based tools that can sift through prior art more efficiently than ever before. This means quicker turnaround times and more accurate assessments. Some countries are already implementing AI to help with initial screenings, setting the stage for a quicker and more reliable review process globally.

Even after a patent is granted, AI continues to play a vital role in monitoring and enforcement. Infringement detection has traditionally been a challenging and resource-intensive task. AI algorithms can scour through mountains of data, identifying potential violations and alerting patent holders in real-time. This capability is especially useful in industries where rapid imitation is common. Companies can protect their intellectual property more effectively, bolstering their competitive edge.

Moreover, AI is democratizing access to patent data. Earlier, access to comprehensive patent databases and analysis tools was a luxury enjoyed only by large corporations or specialized legal firms. Today, AI-powered platforms offer subscription-based services that small businesses and independent inventors can afford. This level playing field allows innovators from diverse backgrounds to contribute to the pool of human knowledge, spurring a more inclusive wave of innovation.

There's a motivational aspect to AI in patent analysis that shouldn't be overlooked. Knowing that AI can handle several tedious aspects of patenting can be a significant relief for inventors. This lets them focus more on the creative and innovative aspects of their work. They aren't bogged down by the minutiae of documentation and legal jargon. The less time inventors spend on administrative tasks, the more time they have to innovate, fueling a virtuous cycle of creativity and invention.

In terms of global impact, AI is standardizing patent practices across different jurisdictions. Different countries often have varying norms and expectations for patent filings. AI tools can adapt to these regional differences, making the patenting process more streamlined on a global scale. This is particularly beneficial for multinational corporations that need to file patents in multiple countries simultaneously.

The inspirational stories abound. Take for instance a small startup that leverages AI to identify a niche in renewable energy technology. By analyzing extensive patent databases, the AI shows that an area within solar energy hasn't been fully explored. The startup quickly files a patent and begins developing technology that could revolutionize the energy sector. This isn't a hypothetical scenario but a reality unfolding in multiple sectors worldwide.

Some might argue that the proliferation of AI in the patent domain could lead to an over-saturation of patent applications. However, it's equally plausible that the enhanced capabilities brought by AI will lead to higher quality patents rather than sheer quantity. AI's analytical skills, combined with human creativity, can refine ideas to their best possible versions before they are even submitted, raising the overall bar of innovation.

AI-driven patent analysis tools are not just confined to the tech sector. Industries as diverse as pharmaceuticals, agriculture, and even fashion are beginning to realize the benefits. The common denominator is the desire to innovate faster, more efficiently, and with greater precision. Imagine a fashion designer using AI to analyze decades-worth of designs to ensure their new collection is entirely revolutionary. They can see what's been done and what remains unexplored, leading to truly original creations.

With all these advancements, it's natural to feel both excited and a bit overwhelmed. The landscape is evolving rapidly, and the rules of the game are changing. However, it's clear that AI's role in patent analysis is a game changer. Companies and individuals who adapt and leverage these tools will undoubtedly find themselves at the forefront of innovation.

It's a brave new world, one where human ingenuity and artificial intelligence work hand in hand. While AI handles the heavy lifting, we can focus on dreaming bigger and reaching further. We are on the brink of an unprecedented era of innovation, driven by the symbiotic relationship between human and machine.

This is more than just a shift; it's a revolution set to redefine what's possible in the realm of inventions and patents. As AI continues to evolve, its role in patent analysis will only grow more profound, ensuring that our brightest ideas come to fruition faster and more effectively than ever before.

Clearly, the future of patent analysis is bright, brimming with possibilities. The synergy between AI and human innovation promises a future where limitations are minimal, and potential is boundless.

Chapter 17:
Real-World Applications of AI

The transformative power of AI is being realized in real-world applications that span across diverse industries, revolutionizing traditional practices and birthing innovative solutions. From agriculture to finance, AI is optimizing processes, increasing efficiency, and driving growth in unprecedented ways. Imagine self-learning algorithms analyzing seismic activity to predict earthquakes or AI-driven predictive maintenance systems revolutionizing manufacturing by foreseeing equipment failures before they happen. In healthcare, AI is not only aiding in swift and accurate diagnoses but also tailoring personalized treatment plans that consider an individual's unique genetic makeup. Retailers are enhancing customer experiences through AI-powered recommendation engines that understand preferences better than humans can, while smart cities are emerging with AI at their core, ensuring safer and more efficient urban living. The case studies and industry implementations discussed will give you a glimpse into how these intelligent systems are not just futuristic concepts but practical tools making a tangible difference right here, right now.

Case Studies

Case studies serve as vivid illustrations of how AI is reshaping industries and impacting our daily lives. Each example we'll explore draws from different sectors, showcasing AI's versatility, trans-formative power, and potential for driving innovation forward. Let's dive into these real-world scenarios and see what lessons we can glean from them.

Our first case study brings us to the realm of healthcare. Consider the work done by a company called IBM with its AI platform, Watson. Watson for Oncology has been instrumental in revolutionizing cancer treatment protocols. Doctors often find themselves overwhelmed by the sheer volume of medical research and treatment options available. Watson steps in by sifting through mountains of data, including patient records, medical journals, and clinical trial outcomes, to recommend personalized treatment plans. This AI-driven solution doesn't just save time; it increases diagnostic accuracy and optimizes treatment plans, offering patients a fighting chance against a formidable disease.

The realm of transportation offers another compelling case study. Picture autonomous vehicles, zipping through city streets, each equipped with sophisticated AI systems that handle navigation, recognize traffic patterns, and even communicate with each other to avoid accidents. Companies like Tesla have made significant strides in this area. Their Full Self-Driving (FSD) capability aims to make human drivers redundant. Using a combination of camera systems, radar, and AI-based decision-making, these vehicles promise to reduce traffic accidents and improve overall driving efficiency. The ripple effects include reduced congestion, lower emissions, and a fundamentally safer driving experience.

Next, let's turn to the retail sector, where companies like Amazon have been trailblazers. They've developed algorithms for inventory management, customer service, and even cashierless stores. Take Amazon Go, for instance. These are physical stores where you can walk in, pick up what you need, and walk out without ever interacting with a cashier. AI manages the entire process, from tracking what items you've picked up to automatically charging your account as you leave.

It's a glimpse into a future where shopping is frictionless and incredibly efficient.

The entertainment industry provides yet another fascinating case study. Netflix is the gold standard when it comes to using AI for personalized content recommendations. Their algorithm examines user behavior—what you watch, when you watch it, how long you stay engaged—to recommend shows and movies you're likely to enjoy. This personalized approach has not only improved user satisfaction but has also kept Netflix ahead in an intensely competitive market. Furthermore, AI-driven content creation tools are starting to assist in scriptwriting, movie editing, and even special effects, opening new frontiers for creativity.

In the financial sector, companies like JP Morgan are leveraging AI to transform operations. Take their COiN (Contract Intelligence) platform, for example. This system uses machine learning to analyze legal documents and extract key data points, a process that would traditionally require significant human effort. The platform can review thousands of documents in seconds, detecting anomalies and reducing the risk of human error. This improves efficiency and allows human resources to be redirected towards more complex tasks, making financial operations smoother and more reliable.

Education has also been significantly impacted by AI. Consider the case of Coursera, an online education platform that uses AI to personalize learning paths for students. By analyzing how students interact with course material, the platform can offer customized recommendations that help learners focus on areas where they need improvement. This adaptive learning approach ensures that each student receives a tailored educational experience, making learning more effective and engaging.

The energy sector provides another example worth noting. Companies like Google have developed AI to optimize data center operations, resulting in significant energy savings. Their DeepMind AI system has reduced energy usage for cooling by up to 40%. It accomplishes this by analyzing data from thousands of sensors within the data center, making real-time adjustments to cooling systems. This not only lowers operational costs but also contributes to broader sustainability goals.

Speaking of sustainability, let's not overlook AI's role in environmental conservation. Non-profits and research organizations are using AI to monitor wildlife populations and combat illegal activities like poaching. For instance, Conservation Metrics employs machine learning algorithms to analyze audio and visual data from remote sensors in natural habitats. These systems can identify specific animal calls or detect human activity, allowing for timely interventions. It's a significant leap forward in protecting biodiversity and preserving ecosystems.

Next, let's move to agriculture—an industry traditionally resistant to high-tech interventions. However, companies like John Deere have incorporated AI into their machinery to enhance farming practices. Their tractors, equipped with AI and IoT (Internet of Things) technologies, can autonomously plow fields, plant seeds, and even monitor crop health. These "smart tractors" gather data as they operate, providing farmers with insights to optimize yield and resource use. It's agriculture, reimagined for the 21st century.

In real estate, AI is making waves by significantly simplifying the home buying process. Zillow, with its Zestimate feature, uses AI to provide real-time property valuations. By analyzing millions of data points, including location, market trends, and property features, Zillow offers highly accurate pricing estimates. This empowers buyers and sellers with critical information, making real estate transactions more transparent and efficient.

Let's not forget the aviation industry. Predictive maintenance is a game-changer, and companies like GE Aviation are leading the charge. By utilizing AI to analyze data from aircraft sensors, they can predict equipment failures before they occur. This approach not only ensures flight safety but also minimizes downtime and reduces maintenance costs. It's a win-win for airlines and passengers alike.

Lastly, the domain of cybersecurity has seen a remarkable infusion of AI technologies. Firms like Darktrace use AI for threat detection and response. Their system mimics the human immune system, continuously learning and adapting to recognize unauthorized activities. When it detects a threat, it automatically responds to neutralize it, all while providing insights for future protection. This advanced layer of security is crucial in an era where cyber threats are ever-evolving.

These case studies span a diverse array of industries, each highlighting the transformative potential of AI. Whether it's healthcare, transportation, retail, entertainment, or another sector, AI is driving significant advancements. These examples show that the future is already here, powered by intelligent machines that improve efficiency, enhance decision-making, and open new possibilities. It's an inspiring landscape for tech enthusiasts, business leaders, and anyone keen to understand AI's impact.

As we see, the true value of AI lies not just in its raw computational power but in its ability to augment human capabilities, solve complex problems, and foster innovation. Case studies like these provide a roadmap for what's possible and set the stage for the next wave of technological evolution.

What lessons can we draw from these experiences? First, it's evident that integrating AI into existing systems can dramatically improve efficiency and outcomes. Second, human oversight remains crucial; AI complements human intelligence, rather than replacing it entirely. Lastly, continuous learning and adaptation are key—with every deployment, AI systems become smarter, more reliable, and better equipped to handle future challenges.

The journey from concept to practical application is seldom straightforward, but as these case studies show, the rewards can be substantial. As we move ahead, more

Industry Implementations

Artificial intelligence is no longer just a concept from science fiction; it's a reality that's reshaping industries across the globe. From healthcare to transportation, AI's applications are cutting across traditional boundaries and creating new paradigms of efficiency and innovation. But what does this look like in the real world?

In manufacturing, AI-driven automation is taking the industry by storm. Gone are the days when factory floors were solely populated by human workers. Today, robots powered by machine learning algorithms are handling repetitive tasks with pinpoint accuracy, improving not only productivity but also safety. Predictive maintenance is another breakthrough, where AI systems predict equipment failures before they happen, minimizing downtime and extending the lifespan of machines.

Retail is another sector witnessing a seismic shift. Think about personalized shopping experiences. AI algorithms analyze customer data to understand preferences and buying habits, offering tailored recommendations that feel almost eerie in their precision. On top of this, inventory management systems use AI to predict demand for products, ensuring shelves are never empty and overstock is minimized. These advancements don't just improve customer satisfaction; they significantly bolster bottom lines.

In the financial industry, AI's impact is just as profound. Fraud detection systems have become increasingly sophisticated, utilizing complex algorithms to identify unusual patterns that may indicate fraudulent activity. This not only safeguards financial institutions but also instills greater trust among clients. AI is also revolutionizing investment strategies through 'robo-advisors,' which provide personalized financial advice and portfolio management at a fraction of the cost of traditional financial advisors.

Transportation is another frontier where AI is making waves. Autonomous vehicles, while not ubiquitous just yet, are no longer a distant dream. Companies like Tesla and Waymo are leveraging complex algorithms to teach cars how to drive themselves. These advancements promise to reduce accidents caused by human error, improve traffic flow, and render urban commuting more efficient. Beyond cars, smart traffic systems are using real-time data to manage congestion, reroute vehicles, and make public transportation systems more reliable.

The energy sector is experiencing its own AI revolution. Utilities are using AI to predict and manage energy consumption, optimizing the use of renewable resources. AI systems can analyze weather patterns and make real-time adjustments to balance supply and demand, thereby minimizing waste and reducing costs. Predictive models can also forecast equipment failures in power plants, helping to avert blackouts and other disruptions.

Agriculture might not be the first industry that comes to mind when you think of AI, but it's one where these technologies are sowing the seeds of significant change. Precision farming uses AI to analyze data from drones and IoT sensors to monitor crop health, soil quality, and weather conditions. This leads to more efficient use of resources, higher crop yields, and, ultimately, safer food supply chains. Smart irrigation systems reduce water consumption by delivering the right amount of water at the right time, and robotic harvesters can pick fruits and vegetables with a level of accuracy and speed human hands can't match.

The entertainment industry, too, is being transformed. Streaming services like Netflix and Spotify harness AI algorithms to provide users with eerily accurate content recommendations. Not only does this keep viewers and listeners more engaged, but it also provides valuable data to content creators about what audiences want. In gaming, AI is creating non-player characters (NPCs) that are more realistic and responsive, enhancing the overall user experience.

In transportation logistics, AI is optimizing routes and ensuring timely delivery of goods. By analyzing traffic patterns, weather conditions, and driving behavior, companies can streamline their delivery processes. This not only brings down costs but also reduces carbon footprints, aligning with broader sustainability goals.

The healthcare sector's leap into AI merits special attention. Diagnostic tools employing AI can evaluate medical images faster and more accurately than human doctors. These systems can identify early signs of diseases like cancer, leading to more timely interventions. Personalized medicine, driven by AI, tailors treatments to individual patients based on their genetic profiles and lifestyle choices, resulting in more effective healthcare outcomes.

AI is also leaving its mark in the legal field. Legal research, a time-consuming process, has been made exponentially faster with AI. Algorithms sift through vast amounts of legal documents to find precedents and relevant cases, enabling lawyers to build stronger cases more efficiently. AI can also predict the outcomes of trials, aiding attorneys in crafting strategies and advising clients on the best course of action.

In education, AI systems are laying the groundwork for more personalized learning experiences. Adaptive learning technologies assess students' strengths and weaknesses in real-time, offering customized lesson plans that cater to individual needs. Virtual tutors powered by AI are available round-the-clock, providing students with the assistance they need, when they need it, and enabling a more flexible learning environment.

The public sector isn't immune to AI's transformative power either. Governments worldwide are adopting AI to improve public services. From predictive policing to intelligent traffic management, these technologies promise increased efficiency and better resource allocation. In disaster management, AI systems can predict and track natural disasters, enabling more effective response strategies and minimizing loss of life and property.

AI's influence extends to the realm of creative industries, too. In arts and design, AI algorithms are generating novel pieces of artwork, music, and fashion. These systems analyze vast datasets of artistic works to output creations that are unique yet familiar, challenging traditional notions of creativity. Similarly, in content creation, AI tools are assisting writers and marketers by drafting articles, designing websites, and creating ad campaigns with minimal human intervention.

Even the real estate industry benefits from AI. Advanced algorithms evaluate property prices by analyzing various factors, including location, market trends, and economic indicators, offering more accurate valuations. Chatbots assist potential buyers by answering queries and suggesting properties based on their preferences, streamlining the search process.

Then there are the humanitarian applications of AI. Non-profits and international organizations are leveraging AI to monitor and address global issues like poverty, malnutrition, and climate change. By analyzing data from multiple sources, these systems can identify hotspots, recommend interventions, and monitor the effectiveness of implemented solutions.

Across the board, AI is enhancing the precision, efficiency, and scope of activities in ways that were barely imaginable a decade ago. But with these advancements come challenges and considerations—ethical, economic, and legal—that must be thoughtfully addressed. As industries continue to harness the power of AI, the key will be to strike a balance between innovation and responsibility. Each application serves as a window into a future where human ingenuity and machine intelligence work hand in hand to create solutions that benefit all of humanity.

Ultimately, the real-world implementations of AI are both a testament to human ingenuity and a harbinger of the massive changes yet to come. By understanding these implementations, we can better prepare for and navigate the AI-driven future that's fast approaching.

Chapter 18:
The Ethics of AI Development

As we delve into the ethics of AI development, we're not merely exploring a technological frontier; we're grappling with the essence of what it means to be human in a rapidly evolving world. Accountability looms large—who's responsible when an autonomous system errs? Bias and fairness are not just theoretical concerns; they're tangible realities that can amplify existing societal disparities or carve out new ones. Navigating these ethical waters demands a deep understanding of both the technology and the human condition. It's about ensuring that the incredible power of AI is wielded responsibly, fostering a future where innovation doesn't come at the cost of our values. Balancing progress and principles, we must embed empathy in our algorithms, justice in our datasets, and transparency in our processes. The future impact of AI hinges on these ethical foundations, shaping systems that are not only intelligent but also just and equitable.

Accountability

Accountability in AI development is a topic that demands our immediate attention. As AI systems become more integrated into our daily lives, pinpointing who is responsible when things go wrong becomes increasingly complex. Imagine a scenario where an autonomous vehicle gets into an accident. Who is to be held accountable? Is it the manufacturer, the software developer, or the owner of the vehicle? These are not just theoretical dilemmas; they have real-world implications that affect lives and livelihoods.

The layered nature of AI systems complicates the debate on accountability. An AI system isn't merely a piece of software running on a computer. It often involves intricate networks of algorithms, data inputs, and machine-learning models that evolve over time. Unlike traditional software, where a specific line of code can be easily identified as erroneous, AI systems can produce unexpected outcomes due to the dynamic nature of their learning process. As such, identifying the 'smoking gun' becomes a labyrinthine task.

Top tech companies are grappling with these questions every day, balancing innovation with moral and ethical responsibilities. On one hand, AI can solve incredibly complex problems, pushing the boundaries of what's possible. On the other hand, if these solutions come without a moral compass, we might win the battle but lose the war for ethical integrity. It's essential to establish frameworks that ensure developers and companies can be held accountable for their AI's actions.

Incorporating accountability in AI starts with transparency. If AI systems are to be trusted, they must be explainable. Hidden behind layers of neural networks, their decision-making process often appears as a 'black box' to even seasoned developers. But this opacity is a double-edged sword. It enhances the allure of AI but simultaneously makes it harder to diagnose faults. For an AI-driven future, transparency isn't just a feature; it's a necessity. Explainable AI, where the system can articulate its reasoning in human-understandable terms, can bridge the gap between complexity and accountability.

Moreover, regulations can play a pivotal role. Just as regulations apply to other highly influential industries like finance and healthcare, the AI landscape needs its own governance structure. While regulations may initially slow down innovation, they bring long-term benefits by ensuring responsible development and deployment. Regulatory bodies could mandate periodic audits of AI systems to check for biases, evaluate decision-making paths, and review data privacy practices. Industries that rely heavily on AI should welcome this oversight as an opportunity to build public trust and credibility.

Additionally, we cannot overlook the role of corporate culture. Companies must instill values that prioritize ethical considerations right from the brainstorming phase of a project. Engineers and developers should be encouraged to ask difficult questions and voice concerns without fearing retaliation. When a problem is identified at an early stage, it's easier and cheaper to address. A culture that promotes ethical behavior and transparency can act as the first line of defense against questionable AI practices.

We must also think globally. In an interconnected world, what happens in one country can have ripple effects across continents. International collaboration in setting accountability standards will be crucial. This global perspective ensures that accountability isn't siloed, making it easier to address issues that cross borders. Think of cybersecurity threats that leverage AI; a collective effort is often needed to counter these effectively.

The challenge with accountability isn't solely in assigning blame, but in creating systems that learn from their mistakes. Preventing the recurrence of errors can be immensely valuable. Just as aviation safety has improved through meticulous accident investigations and the implementation of corrective actions, the same methodology can apply to AI. Post-incident analyses should be integral to AI development, bolstering the system against future failures.

Consider the idea of a "black box" for AI. Much like flight recorders in airplanes, these could log detailed records of AI decisions and actions. Such data can be invaluable for post-event analysis, shedding light on system behaviors and informing developers about potential weaknesses. Introducing such technological accounting measures can help bridge the gap between innovation and safety.

Education and awareness also play significant roles. Developers should undergo training that covers both technical and ethical aspects of AI. This dual focus ensures that they are not only adept at designing functional systems but also cognizant of the ethical ramifications of their creations. A well-rounded educational approach produces professionals who can navigate the complexity of developing responsible AI.

Accountability in AI isn't just about pointing fingers; it's about building resilient systems and responsible practices. Consider this a call to action for tech creators, regulators, and users alike. As laypersons, it's essential to ask critical questions about the AI systems we interact with. Demanding more transparency and ethical behavior from companies can drive changes from the grassroots level. When users champion accountability, it compels developers to strive for higher standards.

While we can strive for perfection, we must also recognize that errors are part of the learning process. Mistakes will happen, but their impact can be mitigated. The focal point shouldn't be on eliminating every mistake but on learning from them effectively. Companies and developers should take a proactive stance, preparing for errors by building robust, fail-safe measures.

The ultimate goal is a balanced approach combining innovation with accountability. If we manage this balance, the possibilities of AI can be explored without sacrificing ethical considerations. As we stand at the cusp of an AI-driven future, accountability isn't just a buzzword; it's the core principle that will guide us towards responsible and meaningful technological advancements.

Bias and Fairness

Bias and fairness are critical considerations in the ethics of AI development. While AI holds the promise to revolutionize countless industries, from healthcare to transportation, the path is strewn with complex moral and ethical dilemmas. One of the most pressing issues is the inherent bias that can seep into AI systems. Left unchecked, these biases can reinforce societal prejudices, erode trust in AI technologies, and result in discriminatory outcomes.

So why does bias happen in AI? A significant factor is the data used to train these systems. Machine learning models rely on vast datasets to function. If these datasets contain biased information—whether due to historical inequities or unrepresentative sampling—the resulting AI will also be biased. For instance, a facial recognition system trained primarily on images of light-skinned individuals will underperform or make more errors when presented with darker-skinned faces. This leads to unfair treatment and could have severe consequences, especially when the technology is used in critical areas like law enforcement.

The issue of bias extends far beyond facial recognition. Consider AI in hiring processes: if the training data includes biases present in past hiring decisions, the AI may perpetuate those biases, leading to unfair hiring practices. This can undermine efforts to diversify workplaces and reinforce existing inequalities.

But data isn't the only culprit. The design and implementation stages also play crucial roles. The developers' and engineers' subjective choices—no matter how well-intentioned—can introduce bias into AI algorithms. Factors like feature selection, data labeling, and even algorithmic design can embed bias into the final product. Sometimes, these biases are unintended. However, the impacts are real and far-reaching.

Addressing bias requires a multi-faceted approach. One strategy involves diversifying the teams behind AI development. Diverse teams bring varied perspectives and are better equipped to identify and mitigate potential biases. This isn't just about ethical correctness; it also makes good business sense. A diverse team can build more inclusive, market-friendly solutions. By fostering inclusivity and collaboration, organizations can create AI systems that better serve all segments of society.

Another crucial tactic is cultivating greater transparency in AI development. When algorithms and data sources are transparent, they become easier to scrutinize and improve. Open models allow third parties to identify biases and propose corrections. This transparency fosters trust among users, who can see how decisions are made and understand that the algorithms are constantly being refined to ensure fairness.

Let's not forget the importance of continuous monitoring. Bias isn't a problem you solve once; it's a persistent risk that requires constant vigilance. Tools and methodologies for auditing AI systems should be part of any comprehensive strategy to mitigate bias. Regular audits can reveal new biases that appear over time, ensuring that the system remains fair and equitable. The ethical goal is to ensure that the technology evolves in a way that supports all users, regardless of background.

One encouraging development in this arena is the rise of fairness metrics and guidelines. Organizations like the Fairness, Accountability, and Transparency in Machine Learning (FAT/ML) community are pioneering frameworks for evaluating and mitigating bias. These guidelines offer practical steps for developers and stakeholders to follow, from the initial design stages through deployment and maintenance. Adopting these standards can significantly reduce the risk of bias creeping into AI systems.

Let's also consider the societal implications. An AI system's unfairness can have devastating real-world consequences. For instance, biased healthcare algorithms might offer less favorable treatment options to certain racial or ethnic groups, exacerbating existing health disparities. Biased financial algorithms might unfairly deny loans to certain demographics, perpetuating economic inequalities. These are not hypothetical scenarios; they're documented cases that illustrate the high stakes involved.

The ethical dimension of bias in AI isn't just about preventing negative outcomes; it's also about fostering positive ones. Fair AI systems can help dismantle existing barriers and offer opportunities for marginalized communities. When AI is leveraged correctly, it can level the playing field and contribute to a more equitable society. Whether it's by identifying disparities in healthcare or offering personalized education solutions to underserved students, fair AI has the potential to create positive societal impacts.

But recognizing the potential isn't enough. Businesses, too, have a vested interest in ensuring their AI systems are fair and unbiased. Companies deploying AI solutions that are perceived as biased can suffer reputational damage and lose consumer trust. Case in point: tech companies that have faced public backlash over biased algorithms have seen significant hits to their brand value. In a world where consumers are increasingly aware of and vocal about social justice issues, fairness in AI is a business imperative.

Moreover, the regulatory landscape is evolving. Governments worldwide are beginning to implement policies aimed at ensuring fairness in AI systems. These regulations are likely to become more stringent, and companies that proactively address AI bias will be better positioned to comply. Staying ahead of regulatory expectations not only mitigates risk but can also serve as a competitive advantage.

From a strategic standpoint, embedding fairness into the DNA of AI development can unlock new opportunities. Consider AI-driven customer service solutions that understand and respect cultural nuances, potentially opening up new markets. Or think about personalized learning platforms that fairly assess students from diverse backgrounds, paving the way for educational innovations. Fairness isn't just the right thing to do; it's a catalyst for innovation.

While the road to fair and unbiased AI is fraught with challenges, it's a journey worth undertaking. The success of this endeavor hinges on collective effort—bringing together academia, industry, and policymakers to formulate robust, ethical AI guidelines. Collaborative research and shared best practices can pave the way for more equitable AI ecosystems.

Let's acknowledge that no system can ever be entirely free of bias. Human biases will always find a way to manifest in the tools we create. However, by striving for fairness and implementing rigorous checks and balances, we can significantly mitigate these biases and create more just systems.

In conclusion, bias and fairness in AI aren't just technical challenges; they are moral imperatives. As we continue to develop and deploy AI technologies, we must prioritize ethical considerations to ensure these powerful tools serve the greater good. By doing so, we can harness the transformative potential of AI in a way that benefits all of humanity, creating a future that is not only technologically advanced but also deeply just and equitable.

Chapter 19:
The Regulatory Landscape

As we transition into an era increasingly defined by artificial intelligence, navigating the regulatory landscape becomes both a challenge and an opportunity. Current laws around AI are a patchwork quilt, stitched together with pieces of legislation from various domains—data privacy, intellectual property, and even consumer protection. We've already seen attempts by governments to harmonize these efforts through guidelines and frameworks, but the pace of innovation often outstrips regulatory adaptability. Future policy considerations must balance fostering innovation with safeguarding public interests, ensuring AI's profound benefits aren't marred by ethical pitfalls or malicious use. Imagine a world where AI systems are as rigorously tested as pharmaceuticals before mass deployment, where algorithmic transparency and accountability aren't just buzzwords but mandated standards. In such a dynamic environment, getting ahead of the curve means staying informed and adaptable, embracing the collaborative efforts between policymakers, technologists, and the public to craft laws that will shape the ethical deployment of AI for years to come.

Existing Laws

To make sense of the current legal environment surrounding artificial intelligence, we need to dig into the existing laws that govern this rapidly evolving field. In many ways, AI has outpaced the regulatory frameworks meant to oversee it. Even so, several key pieces of legislation and regulatory bodies provide a foundational legal landscape that shapes how AI technologies are developed and used today.

One of the most influential frameworks impacting AI is the General Data Protection Regulation (GDPR) enacted by the European Union. GDPR has set a high bar for data protection and privacy, affecting businesses worldwide. The regulation mandates transparency, accountability, and user consent regarding personal data, all of which have significant implications for AI systems that often rely on vast datasets. From predictive analytics to personalized marketing, GDPR ensures that data handling practices are not only ethical but also legally compliant.

The United States, unlike the EU, lacks a comprehensive federal law that addresses AI directly. Instead, it relies on a patchwork of regulations and guidelines from various agencies. The Federal Trade Commission (FTC), for example, has a mandate to protect consumers and enforce antitrust laws, giving it a crucial role in overseeing AI-driven business practices. Meanwhile, the Food and Drug Administration (FDA) monitors AI applications in healthcare, especially concerning diagnostic tools and personalized medicine. The National Institute of Standards and Technology (NIST) is also significant, developing guidelines and standards to ensure the trustworthiness of AI systems.

On the other side of the globe, China is racing ahead with its own AI regulations. The nation's approach is top-down, reflecting its centralized political structure. The Chinese government has implemented a series of strategic initiatives and regulations aimed at fostering AI innovation while safeguarding national security. For instance, the New Generation Artificial Intelligence Development Plan outlines comprehensive goals and guidelines, emphasizing both technological advancement and ethical considerations. The Cyberspace Administration of China also imposes strict data localization laws which impact how AI services operate within the country.

Canada also emphasizes the importance of ethnicity in AI endeavors. The Algorithmic Impact Assessment (AIA), part of the Directive on Automated Decision-Making, ensures that any automated decision system used by the federal government operates under principles of fairness, accountability, and transparency. The directive mandates impact assessments, public disclosures, and ensures human oversight, aiming to minimize the risk of bias and discrimination in AI-driven decisions. This layered approach reflects an emerging trend toward thoughtful, ethical AI governance.

Latin American countries are catching up in terms of AI legislation. Brazil recently enacted its own General Data Protection Law (LGPD), which closely mirrors the GDPR. Other countries in the region are following suit, recognizing the necessity to harmonize AI practices within global standards. This alignment enhances cross-border data flows and AI collaborations, ensuring that AI solutions are compatible with ethical and legal norms internationally.

India, with its digital and technological ambitions, is also shaping its AI regulations. The Personal Data Protection Bill, yet to be enacted, promises to create a robust privacy framework akin to GDPR. In the meantime, the NITI Aayog's national strategy for AI focuses on leveraging AI for inclusive growth, emphasizing areas such as agriculture, healthcare, and education, while also considering the ethical dimensions and societal impact of AI technologies.

A fascinating development in Japan revolves around the interplay of AI and robotics, influenced by the country's cultural acceptance of robots. The AI and robot-related policy framework aims at both promoting innovation and addressing societal challenges such as an aging population. Legislation concerning AI in Japan includes the promotion of the Sharing Economy, ethical guidelines for AI development, and data protection laws comparable to GDPR, ensuring a balanced approach toward technological advancement and ethical responsibility.

Australia, too, has jumped into the regulatory game with its Artificial Intelligence Ethics Framework. This guideline encourages businesses to adopt ethical principles throughout the AI development lifecycle. Additionally, the Office of the Australian Information Commissioner enforces data privacy laws that regulate how personal data is collected, stored, and used, ensuring that AI technologies operate within predefined ethical and legal boundaries.

Globally, international organizations are increasingly stepping into the regulatory space. The United Nations, for example, has initiated discussions around creating a cohesive global AI strategy, emphasizing human rights, fairness, and accountability. The Organisation for Economic Co-operation and Development (OECD) has also published AI principles, focusing on promoting trustworthy AI while fostering innovation. These universal guidelines aim to create a common ground for countries to follow, enhancing international cooperation and consistency in AI regulation.

Even with these efforts, regulatory gaps and inconsistencies continue to pose challenges. The rapid pace of AI advancement often leaves regulators scrambling to keep up. Issues like AI bias, transparency, accountability, and data protection require ongoing scrutiny and responsive legal frameworks. Multinational companies face a labyrinth of regulations that can complicate cross-border AI deployment. Navigating this multifaceted landscape requires a deep understanding of both local and international laws, as well as a proactive approach to compliance and ethics.

Developers and businesses can't afford to overlook the importance of regulatory awareness. With penalties for non-compliance ranging from fines to reputational damage, staying informed about existing laws is crucial. It's not just about adhering to legal requirements; it's also about fostering trust with end-users and stakeholders. As AI systems become more pervasive, maintaining public confidence is essential for sustainable growth.

What lies ahead for AI regulation is uncertain, but what remains clear is that the existing laws offer a necessary, albeit imperfect, foundation. They represent the interplay between innovation and caution, pushing developers to be more conscientious and responsible in their approach. As you navigate through the complex regulatory landscape, remember that this rigorous framework is not a hindrance but a guide to building trust, fostering innovation, and ensuring that AI technologies serve the greater good.

Future Policy Considerations

As we stand at the precipice of an AI-driven future, the need for robust and forward-thinking policies is more evident than ever. Regulatory frameworks will have to evolve at a pace that matches the rapid technological advancements, ensuring that innovation is nurtured while societal and ethical standards are upheld. Can we really anticipate the myriad ways in which AI will influence our lives and the complexities it will introduce? Absolutely, but it requires a comprehensive and adaptive approach to policy-making.

One significant aspect to consider is the framework for AI-centric intellectual property (IP). With AI systems capable of generating novel ideas, artwork, and even scientific discoveries, determining ownership and IP rights becomes a complex challenge. Will the traditional notions of copyright and patents suffice, or do we need an entirely new paradigm to handle such innovations? Future policies must navigate these waters carefully, balancing the interests of creators, developers, and the broader public.

Moreover, AI's impact on employment cannot be overstated. While AI offers the promise of increased efficiency and the elimination of mundane tasks, it also poses the threat of job displacement. How can policy address this dual-edged sword? One approach could be to invest heavily in retraining programs, creating pathways for workers to transition into new roles that AI cannot easily replicate. Another approach might involve economic safety nets and universal basic income schemes. Neither solution is perfect, but both are worth exploring.

We also need to consider the ethical dimensions of AI deployment. Imagine autonomous systems making life-and-death decisions, such as in healthcare or military applications. The ethical frameworks guiding these systems need to be rigorous, transparent, and subject to continuous review. Policies must mandate the inclusion of diverse ethical viewpoints in the development and deployment of AI systems, ensuring that they reflect a broad spectrum of human values.

When it comes to AI in consumer protection, ensuring transparency and accountability in AI-driven decisions becomes pivotal. Consumers must have the right to understand how AI systems arrive at their conclusions, whether it's a loan approval or a hiring decision. Policies could mandate explainability standards, obligating companies to provide clear and comprehensible explanations of AI-driven decisions. Such measures would not only build trust but also facilitate greater accountability.

Data privacy is another critical area where future policy considerations are essential. With AI systems collecting and analyzing vast amounts of personal data, stringent regulations are needed to protect individuals' privacy rights. This might involve updating existing data protection laws like GDPR (General Data Protection

Regulation) and CCPA (California Consumer Privacy Act) to address AI-specific challenges. Policies could also focus on ensuring that individuals have control over their data, including the right to opt-out of AI-driven data processing.

International collaboration is another crucial factor. AI does not respect national borders, and its regulation requires a global perspective. Collaborative policies between nations, as well as international regulatory bodies, could establish common standards and practices. This would not only facilitate smoother trade and innovation but also help in addressing global ethical and security concerns.

The environmental impact of AI is an often-overlooked aspect that should feature prominently in future policies. AI technologies, particularly those involving large-scale data centers and computational power, have significant energy footprints. Regulatory frameworks should incentivize energy-efficient AI practices, perhaps through tax breaks for companies that adopt sustainable technologies or penalties for those that don't.

Another pivotal area of consideration is the standardization of AI technologies. Industry-wide standards could help ensure the interoperability and reliability of AI systems across different sectors. These standards would need to be developed collaboratively by governments, industry leaders, and academic institutions. Policies promoting standardization would not only enhance the quality and safety of AI technologies but also bolster public trust.

Policies focusing on the ethical use of AI in sensitive sectors like healthcare, law enforcement, and social services are essential. These policies must ensure that AI is used to enhance, rather than undermine, human dignity and rights. They could mandate regular audits and impact assessments to evaluate the social and ethical implications of AI systems, ensuring they are aligned with human-centric values and principles.

Addressing AI bias is another significant policy issue. Training data sets used for AI development often reflect existing societal biases, leading to biased outcomes. Future policies should mandate the use of diverse and representative data sets during AI training, as well as regular audits to identify and mitigate biases in AI systems. Such measures would ensure that AI technologies promote fairness and equality.

In terms of cybersecurity, policies must address the vulnerabilities posed by AI systems, particularly in critical infrastructure sectors like energy, transportation, and finance. Future regulations could mandate stringent cybersecurity protocols and regular vulnerability assessments to protect these sectors from AI-driven threats. Collaboration between public and private sectors would be crucial in developing and implementing these protocols.

Engaging with public opinion is another vital policy consideration. As AI continues to advance, public perception and acceptance of these technologies will play a crucial role in their adoption. Policymakers should prioritize transparency and public engagement, involving citizens in discussions about AI's role and impact on society. This could include public consultations, educational campaigns, and transparent decision-making processes.

Policies should also address the accountability of AI developers and users. Clear guidelines must be established to determine who is responsible for the actions and decisions of AI systems. Whether it's developers, companies, or end-users, policies must delineate accountability to ensure that ethical and legal standards are upheld.

Transparency in AI development is another key aspect to consider. Policies could mandate that companies disclose crucial information about their AI systems, including data sources, algorithms used, and decision-making processes. This transparency would allow for independent audits and assessments, ensuring that AI systems operate fairly and ethically.

In conclusion, future policy considerations in the regulatory landscape of AI are multifaceted and complex. They require a nuanced approach that balances innovation with ethical standards, privacy with transparency, and national interests with global collaboration. As we move forward, the guiding principle should be to ensure that AI technologies serve the broader interests of humanity, fostering a future where technological advancement and societal values go hand in hand.

Chapter 20:
Investing in AI

Investing in AI is no longer a speculative venture; it's a strategic imperative for forward-thinking businesses and savvy individuals. With market analysis revealing exponential growth trends and venture capitalists pouring unprecedented funds into AI startups, the opportunities are ripe for those ready to seize them. Think of AI as the gold rush of the 21st century, where the landscape is being shaped by companies harnessing the power of algorithms and data. This is a realm where innovation accelerates at breakneck speeds, creating industries we can't yet imagine and transforming those we thought were static. For the investor with an eye on the future, AI represents not just a sector but a seismic shift in how we understand markets, competition, and value creation. The companies that lead in artificial intelligence are laying the rails for the future economy, and by aligning your investments with this transformation, you're positioning yourself at the pioneering edge of modern industry. So, whether you're a venture capitalist looking to fund the next AI unicorn or a business leader considering the transformative benefits of AI integration, the imperative is clear: now is the time to invest in AI, and in doing so, invest in the future itself.

Market Analysis

We stand on the edge of an unprecedented technological revolution driven by artificial intelligence (AI). The conversation isn't about whether AI will penetrate various sectors but rather how deeply and expansively it will transform them. Investors, business leaders, and tech enthusiasts are all eager to dissect the AI market to carve out their piece of this burgeoning pie.

First, let's look at the sheer scale of the AI market. According to recent market research, the AI industry is projected to grow from $58.3 billion in 2021 to $309.6 billion by 2026. That's a compound annual growth rate (CAGR) of about 39.7%, an almost dizzying pace. Behind these numbers is a profound and palpable excitement that's fueling investment activities. Venture capital firms are pouring unprecedented sums of money into AI startups, some of which could become the titans of their respective industries.

Importantly, AI's growth isn't confined to the Silicon Valley bubble. Europe, Asia, and other regions are doubling down on innovation. Take China, for instance, which has been strategically positioning itself as a global leader in AI. Government initiatives and private sector ventures are collectively boosting China's ambitions to dominate this space by 2030.

AI technologies are not a monolithic block. They span a range of functionalities—from machine learning algorithms optimizing supply chains to natural language processing enhancing customer service. Each of these segments has unique market dynamics and growth trajectories. For instance, the machine learning segment alone attracted $14.9 billion in investments in 2020. The reason? Its versatility; machine learning can be applied to industries as diverse as finance, healthcare, transportation, and entertainment.

Let's zoom in on a couple of sectors. Healthcare's adoption of AI is particularly explosive. Diagnostic tools, powered by AI, are setting new gold standards for speed and accuracy. Think about how an AI algorithm can scan thousands of radiology images in seconds, flagging abnormalities for human review. The impact is twofold: improved patient outcomes and reduced operational costs. Financial analysts predict the AI healthcare market will exceed $45 billion by 2026.

Financial services are another fertile ground for AI. The AI algorithms are already making significant inroads in fraud detection, risk management, and algorithmic trading. Predictive analytics has turned into a cornerstone for investment strategies. The key takeaway for investors here is clear: financial services will continually demand more sophisticated AI solutions, ensuring a stable growth market for years to come.

What about the darlings of the tech world, startups? They're essential in understanding the pulse of the AI market. Startups are uniquely positioned to drive innovative solutions quickly. Unlike established firms with entrenched systems, startups can pivot rapidly, adopting the latest advancements in real-time. Investors are well aware of this, and over the last few years, we've seen a surge in seed and Series A funding directed towards AI-focused startups.

The market isn't without its challenges, though. One significant barrier to entry is the high cost of AI development and implementtation. Training state-of-the-art machine learning models requires substantial computational resources and engineering talent. However, with advancements in cloud computing, the cost curve is expected to bend, enabling smaller players to engage more robustly in the AI landscape.

Moreover, there are regulatory hurdles that could impact market trajectory. As governments grapple with the ethical, privacy, and security implications of AI, regulations and guidelines are being shaped. Businesses and investors alike need to stay tuned to these evolving frameworks as they could dramatically influence market conditions.

Understanding current trends helps, but anticipating future shifts can be an even more powerful way to navigate the AI market. It's crucial to follow industry reports, analyst predictions, and the latest research publications to stay ahead of the curve. Think about quantum computing – once a theoretical dream, now a budding reality. Integrating quantum computing with AI could result in computational breakthroughs, thereby transforming entire sectors overnight.

Another trend to stay attuned to is the democratization of AI. The rise of open-source AI platforms and accessible machine learning libraries is lowering the knowledge barrier for many aspirant AI developers and companies. This is a double-edged sword: while it enables rapid innovation, it also increases competition. It's a fertile field for investors to spot but also to navigate cautiously.

While global perceptions of AI are largely favorable, they're not uniformly enthusiastic. Differing public attitudes across cultures could influence market growth. In some regions, the skeptical public sentiment toward AI, fueled by concerns around job displacement and ethical misuse, might dampen its adoption. Investors must consider these socio-cultural factors when analyzing market potentials in diverse geographies.

Of course, every potential investment comes with its own set of risks and it's no different with AI. The pace of technological advancement is breakneck, and technologies that seemed revolutionary a couple of years ago might quickly become outdated. Investing in AI requires not just financial acumen but also a thorough understanding of technological trends, competitive landscapes, and regulatory environments.

Strategic collaborations and partnerships will also play a pivotal role in accelerating market growth. Major corporations are increasingly entering into collaborations with AI startups to combine innovative talent with extensive resources. Such synergies can exponentially speed up time-to-market for new AI solutions, an attractive proposition for any investor.

Lastly, let's not overlook AI's role in driving market efficiencies. From inventory management in retail to predictive maintenance in manufacturing, the adoption of AI can lead to cost savings and increased operational efficiencies. Investors who recognize these subtleties can place more informed bets, ensuring they're part of the sectors where AI is set to make the most profound impact.

In conclusion, the AI market is rich with opportunities, characterized by rapid growth and multifaceted applications. However, investors need to be discerning, strategic, and adaptable. By understanding market dynamics, staying abreast of technological advancements, and considering socio-cultural factors, they can position themselves to reap the rewards of this AI-driven future.

Venture Capital

If you're looking at the future of AI, you can't ignore the elephant in the room: venture capital (VC). This arena has become the undisputed engine driving the AI innovation train. The investment landscape for AI has been evolving quickly, with VCs playing a pivotal role in molding its course. Whether it's startups developing transformative technologies or established firms pushing the boundaries, venture capital is the lifeblood that fuels these ambitious visions.

Let's break it down. Venture capitalists aren't just throwing money at AI companies—they're making strategic bets on the future. These investors are looking for groundbreaking advancements that promise to disrupt industries, find solutions to existing problems, and even create entirely new markets. The stakes are high, but so are the potential rewards. VCs often look for those 'unicorn' startups—

companies valued at over a billion dollars—that offer scalable solutions and robust business models.

Many people imagine a boardroom full of stern-faced investors making decisions based strictly on numbers and projections. While financial metrics are crucial, venture capitalists are increasingly considering the broader implications and ethical aspects of AI technologies. They're interested in how these innovations impact society, the potential for job creation or displacement, and the ethical frameworks within which these technologies operate. Investors today understand that a product's social impact can significantly affect its market success.

One of the most captivating aspects of venture capital in AI is its inherent unpredictability. In the rapidly changing world of technology, today's breakthrough might be tomorrow's obsolete idea. Hence, diversification is key. VC firms often spread their investments across a range of AI sectors—from healthcare and transportation to cybersecurity and entertainment. This strategy helps them mitigate risks while increasing the chances of hitting upon a groundbreaking innovation.

So, what trends are currently catching the eye of venture capitalists? Machine learning and deep learning, for sure. But there's more. The intersection of AI with other technologies like blockchain, quantum computing, and biotechnology is creating new investment opportunities. Venture capitalists are closely watching sectors like personalized medicine, autonomous vehicles, and AI-driven financial services. Each of these niches offers a unique set of challenges and rewards, making them highly attractive for investment.

Yet, it's not just about spotting the next big thing. Building a successful AI company requires more than just funding. VCs often provide startups with mentorship, strategic guidance, and access to an invaluable network of contacts. This support system can be crucial during the risky early stages of a startup's life, where one wrong move can spell disaster. Venture capitalists thus become integral partners, sharing the entrepreneurial journey toward bringing disruptive AI technologies to market.

You might wonder, what's the risk? Well, venture capital is not for the faint-hearted. The volatile nature of tech development means some investments will inevitably fail. Companies working on the bleeding edge of AI often face substantial technical hurdles, regulatory challenges, and market resistance. Despite thorough vetting processes, it's not uncommon for startups to underperform or pivot drastically from their original business plans.

Then there's the ethical dimension. As AI systems become more integrated into daily life, questions surrounding privacy, bias, and accountability gain urgency. Venture capitalists have to navigate this ethical minefield carefully. There's a growing emphasis on responsible AI—where fairness, transparency, and accountability are not afterthoughts but core principles. Investors are now scrutinizing startups not just for their technological prowess but also for their commitment to ethical standards.

Looking globally, venture capital isn't confined to Silicon Valley. While the Bay Area remains a hotbed for AI innovation, other regions are catching up. Europe, China, and countries like Israel are rapidly emerging as significant players, each with its unique ecosystem and investment trends. This globalization of venture capital is enriching the AI landscape, bringing in diverse perspectives and expertise.

Behind the scenes, data analytics and AI are revolutionizing venture capital itself. Investors are leveraging AI to enhance decision-making, optimize their portfolios, and identify promising startups more efficiently. Predictive analytics, sentiment analysis, and natural language processing tools are becoming part of the VC toolkit, helping investors make better-informed decisions.

Despite the inherent risks and challenges, the promise of transformative AI technology makes venture capital an exciting field. The relationship between VCs and AI startups is symbiotic. While startups need the capital and support to bring their visions to life, VCs rely on these cutting-edge technologies to achieve high returns on their investments.

The landscape will continue evolving, driven by technological advances and shifts in global markets. The success stories of AI-driven companies often traced back to visionary venture capital investments, serve as inspiring reminders of what's possible. For tech enthusiasts and business leaders alike, understanding the dynamics of venture capital in AI isn't just about following the money. It's about grasping the future potential of technologies that could redefine our world.

In essence, venture capitalists are patrons of innovation. They're providing the fuel that keeps the AI engine running, taking calculated risks in the hope of reaping substantial rewards. As we look forward to what's next, one thing is clear: the partnership between AI and venture capital will continue to be a critical force in shaping our future.

Chapter 21:
Preparing for the AI Revolution

A s we stand on the brink of an AI-driven future, the key to thriving in this new era lies in our readiness to adapt and evolve. This chapter isn't just about forecasting changes but also about arming yourself with the skills and strategic foresight needed to navigate them effectively. We'll explore how continuous skill development and proactive strategic planning can position you—and your organization—at the forefront of the AI revolution. It's a call to action, urging tech enthusiasts, business leaders, and forward-thinkers alike to embrace change, foster innovation, and cultivate a mindset of adaptability. By understanding the imminent transformations and preparing accordingly, we can not only mitigate potential disruptions but also seize unprecedented opportunities, ensuring that the AI revolution propels us toward a prosperous and enlightened future.

Skill Development

As we stand at the brink of unprecedented advancements in artificial intelligence, it's crucial to equip ourselves with the right skills. The AI revolution isn't just about technology; it's a fundamental shift in the way we live, work, and interact with the world. To stay relevant, we need a blend of technical expertise, creativity, and adaptability.

One of the first skills worth developing is programming. While not everyone needs to become a software engineer, understanding the basics of coding can be hugely beneficial. Languages such as Python,

R, and JavaScript are essential when working on machine learning and data analysis projects. These languages form the backbone of many AI systems, making them valuable tools in your skill set.

In addition to programming, a solid grasp of data science is becoming increasingly important. Data is the lifeblood of AI, and knowing how to manipulate, analyze, and draw insights from large datasets can set you apart. Courses in statistics, data mining, and machine learning can provide the foundation needed to navigate this complex field. Universities and online platforms offer numerous resources to get you started.

Understanding AI concepts is another critical area. You don't need to delve into the most complex algorithms right away, but familiarity with terms like neural networks, deep learning, and natural language processing is essential. These concepts form the building blocks of AI applications and having a working knowledge can help bridge the gap between theoretical understanding and practical application.

Soft skills are just as important as technical skills in the AI revolution. Critical thinking and problem-solving abilities can enable you to address the unique challenges posed by AI systems. Often, AI will present solutions or patterns that are not immediately obvious, requiring a nuanced analysis to understand and implement effectively.

Collaboration and communication cannot be understated in the realm of AI. Working with multidisciplinary teams that include data scientists, domain experts, and business strategists is commonplace. The ability to articulate technical concepts to non-technical stake-holders is a skill that will make you invaluable in any project setting.

Creativity is a vital, yet sometimes overlooked, skill in the AI domain. AI can handle a tremendous amount of data and automate processes, but it often falls short in the innovation department. Human beings bring creativity to the table, whether it's in developing new algorithms or imagining entirely new applications for existing technologies. Encourage out-of-the-box thinking and don't be afraid to explore unconventional solutions.

Continuous learning is a mindset you'll have to adopt to keep pace with AI's rapid growth. New advancements are made almost daily, and staying updated requires a commitment to ongoing education. Attend workshops, participate in webinars, and join AI communities to share knowledge and stay informed about the latest trends and technologies.

Another critical area of focus is cybersecurity. As AI systems become more integrated into various aspects of life and business, understanding how to protect these systems from malicious attacks will be indispensable. Knowledge of encryption, security protocols, and ethical hacking can provide a robust defense against potential threats.

One should not overlook ethical considerations in AI development. As AI systems gain more influence, understanding the ethical implications of these technologies becomes paramount. Training in ethics can help you navigate the moral dilemmas that will inevitably arise, ensuring that the solutions you develop are fair, transparent, and accountable.

Investing in leadership skills can also be highly rewarding. As organizations increasingly rely on AI, leaders who understand its capabilities and limitations will be in high demand. Effective leadership can guide teams through the complexities of AI projects, fostering an environment where innovative ideas can flourish.

As you develop these skills, it's also important to apply them in real-world settings. Internships, freelance projects, or collaborative research opportunities can provide invaluable practical experience. This hands-on practice not only solidifies your learning but also helps you build a portfolio that showcases your capabilities to potential employers.

Furthermore, building a robust professional network can significantly enhance your career prospects. Attend conferences, engage in online forums, and connect with professionals in the field. Networking can open doors to opportunities you might not have considered, including job offers, collaborative projects, and mentorship.

Don't underestimate the power of mentoring in skill development. Whether you're the mentor or the mentee, these relationships can provide mutual benefits. Experienced professionals can offer insights and guidance, helping you overcome hurdles that might otherwise slow your progress. Conversely, mentoring others can reinforce your own understanding and present new perspectives on familiar problems.

If you're a business leader or manager, fostering a culture of skill development within your organization can pay immense dividends. Encourage your team to pursue learning opportunities and provide resources for skill enhancement. Promoting a learning culture can drive innovation and adaptability, ensuring your organization stays competitive in the AI-driven market.

Finally, adaptability is perhaps the most crucial skill of all. AI is a fast-moving field, and the ability to pivot and adjust to new information and circumstances is invaluable. Embrace change, maintain a flexible approach, and be open to unlearning outdated methods while learning new ones.

The AI revolution promises vast changes and opportunities. By focusing on skill development—technical prowess, soft skills, ethics, continuous learning, and adaptability—you can not only navigate this transformative era but thrive within it. This journey will undoubtedly require effort and dedication, but the rewards, both personal and professional, are well worth it.

Strategic Planning

The AI revolution isn't just a distant future concept; it's happening right now, reshaping businesses and industries across the globe. Strategic planning for this monumental shift is crucial for anyone who wants to navigate, and more importantly, thrive in this ever-evolving landscape. It's not enough to just understand AI or be aware of its potential applications. The key lies in proactively designing a roadmap that integrates AI into your long-term business strategies.

Successful strategic planning begins with an honest assessment of where you currently stand. This involves identifying which areas of your business are ripe for improvement through AI integration. Are your current processes bogged down by inefficiencies? Is there a lot of repetitive, manual work that could be automated? Or perhaps you have a trove of untapped data that could yield meaningful insights if analyzed by AI algorithms. These questions help pinpoint the starting points for your AI journey.

Balancing immediate needs with long-term goals is another critical aspect. Many companies make the mistake of focusing too much on quick wins—automating a few tasks here and optimizing some processes there. While these are worthy goals, a longer-term vision must be established simultaneously. Think about not just what AI can do for you today, but what transformative impact it can have on your business five or ten years down the line.

Moreover, it's essential to recognize that AI isn't a one-size-fits-all solution. Different industries and sectors will have varied applications and benefits from AI. For example, while predictive analytics might be a game-changer for finance, healthcare could benefit more from diagnostic tools powered by machine learning. Hence, tailor your strategic plans to fit the unique demands and opportunities within your industry. A bespoke approach often yields better results than a generic one.

Also, consider forming cross-functional teams that bring together technology aficionados and business strategists. This fusion of skills ensures that the AI implementation is both technologically sound and aligned with your business objectives. Cross-pollination of ideas between different departments can often lead to innovative solutions that might not have been evident initially.

It's important to cultivate an AI-first culture within your organization. Encourage employees to view AI not as a threat to their jobs, but as a tool that can augment their capabilities. Frequent training sessions, workshops, and open forums can go a long way in demystifying AI and easing apprehensions. When your team is on board and excited about AI, the implementation process becomes a collective effort, significantly increasing the chances of success.

Incorporating AI into your strategic plans also means being prepared for ethical considerations and potential pushbacks. Ethical AI is not just a buzzword; it's a necessity. Make sure your AI systems are transparent, fair, and accountable. This ethical grounding will ensure your AI initiatives are not only successful but also sustainable in the long run. It's easier to build trust with stakeholders, customers, and regulators when your AI systems adhere to a high ethical standard.

Budgeting for AI initiatives deserves special mention. While the initial costs can be substantial, the long-term benefits often far outweigh the expenses. Think of this as an investment rather than a cost. Break down your AI budget into manageable phases, each tied to specific milestones and deliverables. By doing so, you can prove ROI to stakeholders at each stage, which can help secure further funding and resources down the line.

Integration of AI should also include regular audits and performance reviews. AI systems are dynamic and require constant monitoring and adjustments to stay relevant and effective. Periodic audits help ensure your AI initiatives are on track and provide insights into areas where improvements can be made. This iterative process is critical to maintaining a competitive edge in a fast-changing marketplace.

Partnerships and collaborations should not be overlooked. Many companies find success through strategic partnerships with AI vendors, startups, and academic institutions. These collaborations can offer fresh perspectives, access to advanced technologies, and even shared risks. By pooling resources and expertise, you can accelerate your AI projects and broaden their scope.

Additionally, don't underestimate the value of scalability when it comes to AI. As your business grows, your AI solutions should be able to scale along with it. Early on, consider how your AI systems can handle increased data loads, more complex queries, and a broader range of applications. Scalability ensures that your AI investments continue to deliver value as your business evolves.

Risk management is another critical component of AI strategic planning. Despite its transformative potential, AI comes with its own set of risks—data breaches, algorithmic biases, and technological obsolescence, to name a few. It's paramount to build a robust risk management framework that can identify, assess, and mitigate these risks. Effective risk management not only saves you from potential pitfalls but also builds greater resilience in your AI strategy.

Finally, always keep an eye on the future. AI is a rapidly evolving field, and what's cutting-edge today might be obsolete tomorrow. Stay updated with the latest advancements, trends, and best practices in AI. Participate in industry conferences, join AI-focused networks, and regularly consult with experts to ensure your strategic plans remain informed and relevant.

Strategic planning for the AI revolution is not just a task—it's an ongoing journey. It requires foresight, adaptability, and a proactive mindset. With the right strategies in place, you can turn the AI revolution from a disruptive force into a powerful ally for your business, driving unprecedented levels of growth and innovation. The future is AI, and with thoughtful planning, it can be a future where your organization thrives.

Chapter 22:
Expert Opinions on AI

When discussing the transformative potential of artificial intelligence, expert opinions are a treasure trove of insights and foresight. From industry pioneers to academic stalwarts, these thought leaders shed light on where AI is headed and its multifaceted impact on different sectors. Some are optimistic, envisioning AI as a catalyst for unprecedented innovations and efficiency, while others caution against unchecked advancements, emphasizing the ethical ramifications and the need for robust regulatory frameworks. What unifies these experts, however, is their shared belief that AI will be an integral part of our future. As we delve into their perspectives, it becomes clear that understanding AI's trajectory requires a balanced view, blending enthusiasm with caution and foresight with pragmatism.

Interviews

One standout conversation was with Dr. Emily Chen, a machine learning expert who's been at the forefront of AI research for over a decade. Dr. Chen highlighted how AI's role is shifting from being a mere tool for data analysis to becoming a co-pilot in decision-making processes across industries. She mentioned, "We are moving towards a future where AI doesn't just suggest options but helps shape the strategy behind those choices." Her insights were both revealing and thought-provoking, especially when she spoke about AI's potential in addressing complex global challenges such as climate change and public health.

Another fascinating discussion was with Raj Patel, a tech entrepreneur who founded an AI startup focused on enhancing customer experiences. Raj stressed the importance of keeping the human element at the center of AI development. "Ultimately, AI should aim to augment human capabilities rather than replace them," he stated. He shared various examples from his own company where AI had been used to significantly improve user satisfaction while reducing operational costs. His perspective reinforces the notion that AI's real value lies in its ability to work in tandem with human ingenuity.

In contrast, Dr. Laura Fernandez, a cognitive scientist, offered a more cautious outlook on AI's rapid advancement. She raised ethical concerns regarding data privacy and the potential abuse of AI technologies. According to Dr. Fernandez, "The developments in AI are exciting but also bring up questions about surveillance and the erosion of personal freedom." This conversation opened up a dialogue on the need for stricter ethical guidelines and regulations to ensure that AI grows in a way that benefits society as a whole. Her words serve as an important reminder that while AI holds great promise, it also comes with substantial responsibility.

Conversations with industry veterans like Michael Lee, a cybersecurity expert, shed light on the increasing importance of AI in national defense. Michael detailed how AI algorithms are being used to predict and mitigate cyber threats, providing a new layer of security in a world that grows more digitally connected every day. "AI is becoming our best line of defense against complex and evolving cyberattacks," he explained. Michael's perspective was particularly eye-opening for business leaders looking to safeguard their digital assets in this age of cyber warfare.

Our interview with Dr. Samantha Williams, an AI ethicist, brought a unique perspective to the table. Dr. Williams has been working on the development of fair and unbiased AI models, and her focus lies in ensuring that AI development is both inclusive and transparent. "Bias in AI can have far-reaching consequences, impacting everything from job applications to judicial decisions," she emphasized. Dr. Williams' commitment to ethical AI brings to light the critical need for developers to actively address bias and strive for fairness in their algorithms.

On a more optimistic note, our conversation with Julian Moore, a venture capitalist who specializes in AI investments, focused on the economic opportunities brought about by AI advancements. Julian expressed confidence in the sustained growth of the AI sector, driven by its vast applications across industries. He noted, "AI is a fertile ground for innovation, and we are only scratching the surface. Investing in AI now is like investing in the internet in the '90s." His bullish outlook on AI as a driver of economic transformation was quite inspiring, especially for those considering diving into the world of AI entrepreneurship.

In talking to Dr. Alice Gupta, a pioneer in AI applications in healthcare, we learned about the groundbreaking work being done to use AI for early diagnosis and personalized treatment plans. Dr. Gupta enthusiastically shared, "AI is revolutionizing healthcare by enabling faster and more accurate diagnoses, which can significantly improve patient outcomes." She provided case studies indicating how AI has already started to make a difference in early cancer detection and chronic disease management, bringing hope to millions around the globe.

To wrap up, our interview series wouldn't be complete without touching on AI's role in education. Dr. Mark Thompson, an education technologist, eloquently described how AI-powered adaptive learning systems are personalizing education like never before. "Every student learns differently, and AI can help tailor educational content to meet individual needs," he remarked. His thoughts on the role of AI in creating a more equitable and efficient education system were particularly motivating for anyone involved in educational reform.

In summary, these interviews offered a window into the minds of those who are not just observing but actively shaping the AI landscape. While their viewpoints varied, a common thread ran through them: AI is a powerful tool that, if harnessed responsibly, has the potential to address some of the most pressing issues facing humanity today. The insights gathered from these conversations are invaluable resources as we navigate the complexities and opportunities that an AI-driven future promises.

Predictive Analysis

Predictive analysis is more than just a trendy buzzword in AI circles; it's the essence of what differentiates advanced AI systems from their basic predecessors. At its core, predictive analysis involves using historical data to make informed predictions about future events. Through this, AI can provide invaluable insights across various domains, from healthcare to business operations.

Imagine a world where your business decisions are guided by precise forecasts updated in real-time. That's the promise of predictive analysis in AI—ushering in a new era where informed decisions aren't just a luxury but a standard. Leaders who harness this power can navigate markets with unparalleled agility, foreseeing shifts before they happen.

One compelling example is the retail industry. Retailers using AI for predictive analysis can anticipate consumer demands, optimize inventory levels, and even devise targeted marketing strategies. This isn't mere speculation; it's a transformation driven by data science. By analyzing purchasing patterns, weather forecasts, social media trends, and even local events, AI can accurately predict product demand for each store, significantly reducing waste and enhancing customer satisfaction.

In healthcare, predictive analysis holds the promise of revolutionizing patient care. Imagine a world where doctors can predict the onset of diseases such as diabetes, cancer, or cardiovascular conditions even before symptoms appear. Preventive measures could be taken, tailored treatments could be developed, and health outcomes could improve dramatically. AI models analyze vast amounts of medical records, genetic information, and lifestyle data to predict patient health trajectories, potentially saving countless lives.

But predictive analysis isn't just about the here and now; it's about weaving a tapestry of potential futures. For business leaders, this means not just reacting to the market but actively shaping it. With AI, they can spot emerging trends and pivot strategies faster than ever. It's almost like having a crystal ball, only this time it's powered by algorithms and data.

AI's role in economic forecasting extends further. Governments and financial institutions leverage predictive models to assess economic health, predict recessions, and stabilize markets. Central banks might use AI to monitor complex economic indicators and propose monetary policies. These models consider a multitude of variables, from unemployment rates to international trade dynamics, creating a holistic economic forecast that's both nuanced and dynamic.

In transportation, predictive analysis and AI make for a trans-formative partnership. Autonomous vehicles don't just navigate—they anticipate. By processing real-time traffic data, weather conditions, and even pedestrian behaviors, these vehicles can predict potential hazards and adjust their routes dynamically. This capability doesn't just promise safer roads; it also signals a future where traffic jams are minimized, and travel becomes more efficient.

The application of predictive analysis in AI isn't without its challenges. Ensuring the accuracy of predictions is paramount, and this requires high-quality data. Incomplete or biased data can lead to flawed predictions, which might cause more harm than good. Moreover, while predictive analysis tools are becoming more sophisticated, they're not infallible. Human oversight remains critical to interpret and validate these predictions.

Collaboration is key. Imagine AI not as a replacement but as a partner—a sophisticated tool that works alongside human insight. This partnership can elevate decision-making processes, blending analytical rigor with human intuition. As we envision the future, the goal isn't to remove humans from the equation but to enhance our capabilities exponentially.

There's an optimistic streak to predictive analysis that should not be overlooked. It hints at a future where knowledge gaps are bridged and opportunities are realized. Businesses can foresee customer needs, healthcare can shift from reactive to preventive, and economies can dodge downturns before they manifest. This isn't science fiction; it's where we're headed.

For entrepreneurs and innovators, predictive analysis offers a playground of opportunities. Startups that effectively harness predictive modeling can disrupt traditional industries. For example, logistics companies can optimize delivery routes in real-time, cutting costs and improving efficiency, all thanks to AI's predictive prowess. Industries ripe for innovation are those with high data availability but low current utilization of predictive tools.

However, ethics and transparency in predictive analysis deserve attention. When AI models decide who gets a loan, what ads you see, or who gets priority in medical treatments, the decision-making process must be transparent and fair. If predictive analysis becomes a black box, where decisions can't be explained or scrutinized, trust erodes quickly. Ensuring these systems are understandable and free from bias is crucial for widespread adoption and acceptance.

There's also the matter of continuous learning. AI systems equipped with predictive analysis are not static; they evolve. With every data point they absorb, they refine their accuracy, continually fine-tuning their models. This dynamic learning process is akin to a feedback loop that keeps enhancing the system's performance, making future predictions even more reliable.

Aspired accuracy in predictions fundamentally changes strategizing. In marketing, for instance, predictive models can segment audiences more precisely than ever. By analyzing behavior patterns, AI can forecast which demographics are most likely to engage and convert, enabling hyper-targeted campaigns. Businesses can thus allocate their marketing budgets more efficiently, yielding higher returns on investment.

In a broader societal context, predictive analysis can play a crucial role in disaster management. Predictive models can forecast natural disasters like hurricanes, earthquakes, or floods, providing early warnings that save lives. Governments can prepare and respond more effectively, minimizing the human and economic toll of such events. This proactive approach not only mitigates risks but also showcases the societal benefits of AI-driven predictive analysis.

The technical underpinnings of predictive analysis involve complex algorithms and massive computational power, but the essence is remarkably straightforward. By recognizing patterns in historical data, AI systems can make educated guesses about the future. These systems employ various statistical techniques, machine learning models, and sometimes even deep learning architectures to achieve their predictions.

In summing up, predictive analysis in AI isn't just about guessing what happens next. It's about creating a future that's informed, proactive, and poised for opportunities. For those steering the ship of enterprise or crafting policies, this tool is indispensable, bridging the gap between uncertainty and strategic foresight.

So, as we navigate this AI-driven era, let's view predictive analysis as a powerful ally. It's a compass that not only points to new territories but also illuminates the paths we hadn't considered. Equipped with this knowledge, the potential to achieve remarkable advancements is immense, and it's a journey worth embarking on.

Chapter 23:
Global Perspectives on AI

From bustling metropolises in Asia to emerging tech hubs in Africa, AI is reshaping the global landscape in ways that are both unprecedented and exciting. Countries around the world are racing not just to adopt but to innovate in AI, creating a dynamic blend of collaboration and competition. Europe is focusing heavily on ethical AI and data privacy, reflecting its broader regulatory environment, while North America is leveraging its robust tech ecosystem to fuel exponential growth in AI startups. Meanwhile, nations like China and India are utilizing AI to address uniquely local challenges, ranging from urban planning to rural healthcare. What's fascinating is the diversity of applications and the shared goal of harnessing AI to improve quality of life and boost economic productivity. Each region's approach is informed by its cultural, economic, and political contexts, crafting a rich, global tapestry of AI advancements. As international collaboration flourishes, we're witnessing the dawn of a collective intelligence that's more interconnected than ever before—promising a future where AI's benefits know no borders.

Worldwide Adoption

Artificial Intelligence is no longer a term relegated to science fiction or academic papers; it's become an integral part of global society. From the bustling streets of Tokyo to the serene villages of rural Kenya, AI is touching lives in ways previously unimaginable. Countries worldwide are not just witnessing the AI revolution but actively participating in it, adopting technologies in sectors as diverse as healthcare, agriculture, and entertainment. The uptake has not been uniform, with some nations leading the charge while others tread more cautiously, creating a complex yet fascinating mosaic of global AI adoption.

Different regions have distinct strategies and priorities when it comes to integrating AI into their frameworks. For instance, in Asia, China has emerged as a front-runner, investing heavily in AI research and development. The Chinese government regards AI as a key driver of its economic and military might. With plans to become a global leader in AI by 2030, China's strategies encompass everything from state-sponsored research to partnerships with private companies. On the other side, Japan focuses on robotics and eldercare, owing to its ageing population. Japan's vision of Society 5.0 aims to blend cyberspace with physical space, representing a holistic and human-centric approach to AI development.

Europe's journey with AI is unique in its own right, driven by a balance between innovation and regulation. The European Union places significant emphasis on ethical considerations and the protection of individual rights, reflecting a cautious yet robust approach. The General Data Protection Regulation (GDPR) has influenced how data-driven AI systems operate within and beyond Europe. Countries like Germany and the UK have established themselves as key players. Germany, leveraging its industrial base, has made substantial strides in implementing AI in manufacturing, while the UK excels in fintech and healthcare applications.

The United States stands as a powerhouse too, home to Silicon Valley and some of the world's largest tech giants like Google, Amazon, and Apple. American AI development is characterized by a vibrant ecosystem of startups, universities, and corporations all striving to out-innovate each other. Federal initiatives, including the AI

Initiative Act, aim to bolster the country's competitive edge, while private sector achievements frequently make headlines. Yet, there are concerns about job displacement and ethical implications, prompting a slew of discussions around regulation and responsible AI.

Meanwhile, India's approach to AI adoption emphasizes inclusivity and leveraging technology to overcome critical societal challenges. Initiatives like AI for All focus on amplifying the impact of AI in agriculture, healthcare, and education, aiming to bridge the socio-economic gaps. With its coding talent and burgeoning startup ecosystem, India has the potential to transform itself into an AI hub. Government policies and frameworks are being tailored to nurture this expansion, with collaborations between public institutions and private enterprises becoming more prevalent.

African nations are also making remarkable strides, albeit at a different pace. Countries like Kenya and Nigeria are focusing on how AI can address local issues such as predictive agriculture, healthcare diagnostics, and financial inclusion. In Kenya, AI-driven solutions are being implemented to forecast crop yields, thereby helping farmers with better farming decisions. Nigerian startups are leveraging AI to improve healthcare accessibility and financial transactions, contributing significantly to the continent's digital economy.

South America, though often underrepresented in global tech narratives, is also embracing AI. Brazil and Argentina are notable examples, with substantial investments in AI-related research and practical applications. In Brazil, AI is being used to tackle societal issues ranging from crime prevention to healthcare. The government and private sector are keen on harnessing AI's potential to spur economic growth and improve the quality of life for its citizens.

The Middle East, particularly the UAE and Saudi Arabia, has also made AI a cornerstone of its future vision. Dubai aims to be one of the world's smartest cities, incorporating AI for everything from traffic management to government services. The UAE has launched several initiatives, including the establishment of the world's first Minister for AI. Meanwhile, Saudi Arabia's Vision 2030 program sees AI as a critical component for economic diversification, focusing on sectors like energy, entertainment, and public services.

In summary, the global landscape of AI adoption is rich and varied, reflecting the diverse priorities and capabilities of different regions. This widespread integration of artificial intelligence holds the promise of profound advancements but also presents a set of challenges. Diverse adoption rates and tactics underscore the need for international collaboration to address ethical, regulatory, and societal issues, ensuring that the AI revolution is as inclusive and beneficial as possible.

AI's global adoption is not merely about deploying technology; it's about rethinking systems and paradigms that have existed for decades. It challenges us to ponder deeply about the ethical dimensions of data usage, privacy, and the potential biases embedded in algorithms. This raises questions about governance, whose answers will shape the future of AI on a worldwide scale.

As we navigate this dynamic landscape, the world will likely witness a shift in geopolitical power structures influenced by AI capacities. Countries that can effectively harness AI's capabilities while mitigating its risks will potentially emerge with substantial strategic advantages. This makes international dialogue paramount, as nations grapple with the dual objectives of innovation and responsibility. It is this balance that will dictate not just the immediate future but the trajectory of AI on a planetary scale.

Innovation hubs and tech incubators are sprouting worldwide, fostering an environment where the next big AI breakthrough could come from any corner of the planet. Facilitated by global networks of collaboration, these centers are breeding grounds for innovative solutions to both local and global challenges. For instance, initiatives like AI for Earth, supported by global entities, encourage sharing knowledge and resources across borders to tackle climate change. Such collaborative efforts exemplify the spirit of global AI adoption, transcending geographical and political boundaries.

The conversation about AI adoption is a part of a much larger discourse about our collective future. Governments, industries, and civil societies must work together, ensuring that AI benefits are equitably distributed. This is particularly crucial for developing nations, which may lack the infrastructure and resources to compete on equal footing. Investments in education, regulatory reforms, and international partnerships can play a pivotal role in leveling the playing field.

Finally, it's worth considering the human element in all this. How we adapt to and interact with AI will define its ultimate impact. As AI systems become more ingrained in daily life, societies will need to develop new skills and mindsets. Learning to coexist with advanced AI will require a paradigm shift in how we think about work, relationships, and even our own potential.

As we stand at this precipice, the global trend towards AI adoption invites us to dream big yet tread cautiously. Our innovative spirit is matched by our responsibility to shape a future where technology amplifies human potential rather than diminishes it. The path forward is not just about machines and algorithms, but about how we, as a global community, choose to navigate and steer this transformative wave.

International Collaboration

As AI continues reshaping the economic, social, and technological landscapes, the case for international collaboration becomes increasingly compelling. Think about it: the promise of AI transcends borders, languages, and even political ideologies. A shared vision could accelerate breakthroughs that don't just benefit one nation but potentially transform human life as we know it. Collaboration is more than just a buzzword here; it's a necessity for moving forward.

The powerhouses of AI research and development—the United States, China, and the European Union—each bring unique strengths to the table. The U.S., for instance, is a leader in software innovation and entrepreneurial spirit. Silicon Valley, the epicenter of tech innovations, has birthed countless startups and tech giants, leading the charge in AI commercialization. China, on the other hand, excels in the sheer scale of its data reserves and government-backed initiatives, propelling its AI prowess forward. The European Union focuses on ethical AI, emphasizing privacy, fairness, and accountability. By pooling these diverse strengths, we can create a balanced ecosystem where technology thrives responsibly.

However, it's more than just pooling resources. International collaboration hinges on mutual trust and transparent sharing of both successes and failures. The spirit of "coopetition"—a blend of cooperation and competition—can drive partners to push boundaries while maintaining global ethical standards. Consider, for example, the Human Genome Project, a monumental feat achieved through international collaboration. The same principles could easily be applied to AI, expediting discoveries that could help solve some of humanity's most pressing challenges, like climate change and pandemics.

But collaboration isn't devoid of challenges. One significant hurdle is the disparity in regulatory frameworks. For instance, the General Data Protection Regulation (GDPR) in Europe and similar regulations are considerably stricter than what exists in other regions. Aligning these different frameworks without stifling innovation is the million-dollar question. Nations must strike a balance, ensuring security and privacy without hampering technological progress.

Moreover, there's the issue of data sovereignty. Countries are increasingly cautious about where their data is stored and who has access to it. This caution isn't unfounded; data is the new oil in the digital economy. Effective international collaboration must navigate these treacherous waters, ensuring that data sharing does not compromise national security.

One fascinating avenue for collaboration is in addressing global talent shortages. AI talent is scarce, and the race to recruit the best minds can sometimes be cutthroat. Yet, international partnerships in education and research could cultivate a new generation of AI experts, well-versed in diverse cultures and perspectives. By establishing AI-focused research institutes and exchange programs, we can foster cross-border intellectual symbiosis. Shared conferences, webinars, and collaborative research projects amplify this synergy, creating a collaborative ecosystem.

While we often celebrate the successes of AI, failures too offer valuable insights. Take, for example, the stumbles of autonomous vehicle (AV) projects across different countries. Sharing these lessons openly can prevent repeated mistakes, allowing for more robust development cycles. Global forums and consortia, such as the Partnership on AI and the Global Partnership on Artificial Intelligence (GPAI), already serve as platforms for such exchanges, enabling a more transparent and inclusive dialogue on AI's future.

Funding opportunities also offer fertile ground for collaboration. Imagine a world where international consortia fund AI projects that tackle climate change, global health, or food security. Such financial partnerships could pool resources from multiple countries to fund larger, more ambitious projects that single nations might find challenging to support alone. Collaborative funding models would catalyze innovations with far-reaching impacts.

Next, let's talk about ethics. With AI systems becoming ubiquitous, ethical considerations become paramount. International collaboration can embody shared ethical frameworks, promoting technologies that reflect collective human values. Ethical AI isn't just a challenge for tech developers; it's a societal requirement. Collaborative efforts could lead to creating universal guidelines that govern AI development, mitigating biases and ensuring fairness.

Consider the benefits of combining local know-how with global expertise. The agricultural sector, for instance, stands to gain significantly from AI innovations. Precision farming techniques developed in tech hubs could be customized to local farming conditions in developing countries, optimizing yields and reducing waste. Farmers worldwide could benefit from a shared pool of knowledge and tools, leveling the playing field in food production.

Nevertheless, we must also be cautious of potential pitfalls. The complexities of intellectual property (IP) rights and data privacy laws vary across jurisdictions. An imbalance in these areas can lead to foul play, stifling the collaborative spirit. Transparent agreements and globally accepted standards can mitigate these risks, ensuring that collaboration remains fair and open.

In an ideal scenario, global collaboration in AI would resemble a well-coordinated orchestra, where each player has a role yet contributes to a harmonious symphony. The inputs—whether they're algorithms, data sets, or ethical considerations—would harmonize to create outputs that resonate universally. The United Nations and other international bodies could act as conductors, ensuring that everyone stays in tune and that the music plays on smoothly.

Furthermore, let's not overlook the untapped potential in developing nations. Countries in Africa, South America, and parts of Asia present unique use-cases and challenges that can spur innovative AI solutions. By including these nations in the AI conversation, we don't just democratize technology; we also uncover novel applications that benefit the global community.

There's a term in the world of machine learning—transfer learning. It's about using a model trained on one task for another related task. A similar concept applies to international collaboration in AI. Insights and innovations from one part of the world can be leveraged and adapted to solve problems elsewhere, creating a cascading effect of benefits across regions. This kind of knowledge transfer is the real magic of international collaboration.

Sure, nationalistic tendencies and geopolitical tensions can complicate collaborative efforts. But the stakes are high enough that setting aside these differences becomes imperative. AI has the potential to solve global challenges like never before. Think of eradicating diseases, mitigating climate change, and ensuring food security—all within our reach. Collective effort is the key to unlocking these monumental achievements.

In summing up, international collaboration in AI isn't just an idealistic vision; it's a practical necessity. The future we're building demands a collective approach, bringing together the best minds, resources, and ethical considerations from around the globe. This shared journey promises a more inclusive, innovative, and equitable future for all. Let's work in harmony, tap into our collective intelligence, and aim for breakthroughs that serve humanity at large. It's no longer a matter of if we should collaborate but how we can do it most effectively and inclusively. Because when we collaborate, the possibilities are limitless.

Chapter 24:
The Role of AI in Future Technologies

A I isn't just the next big thing—it's the transformative force behind what's coming next in technology. Imagine a world where AI collaborates seamlessly with quantum computing, decrypting complexities at speeds we've only dreamt of, or where advanced robotics not only perform tasks but adapt and learn in real-time. AI will be the cornerstone, integrating with and enhancing these futuristic tech arenas, creating ecosystems of innovation. Whether it's reshaping our concepts of what machines can do, or tackling problems previously deemed insurmountable, AI stands central, pushing the envelope of possibility. As we move forward, embracing this symbiosis will not only propel technological evolution but also chart new paths for industries, economies, and societies worldwide.

Quantum Computing

Quantum computing is more than a technical revolution; it's a paradigm shift that promises to reshape the landscape of future technologies, especially when intertwined with AI. Imagine a realm where classical limitations are transcended. Traditional bits, restricted to 0s and 1s, give way to qubits, which can inhabit multiple states simultaneously. This change underpins the immense computational power of quantum machines, opening doors to possibilities that were once deemed science fiction.

With quantum supremacy now within reach, AI stands poised to benefit immensely. Today, AI algorithms often struggle with complex, multifactorial problems requiring significant computational resources. Quantum computers, leveraging principles like superposition and entanglement, can potentially handle these tasks exponentially faster. They might crack intricate problems related to database searching, optimization, and machine learning in seconds, making our current fastest supercomputers look like abacuses in comparison.

In the realm of machine learning, a subset of AI, quantum computing promises to expedite training processes. Currently, developing robust AI models involves consuming vast amounts of data and computing time. Quantum algorithms, such as the quantum version of machine learning and quantum neural networks, could drastically reduce these timescales. The key to this acceleration lies in quantum parallelism, where multiple computations happen concurrently. Imagine training deep learning models in minutes instead of weeks.

But let's take a moment to dwell on the profound philosophical implications. Faster computing power is not just about efficiency; it's about expanding the horizons of what we can understand and achieve. The interplay of quantum computing and AI could offer novel insights into the fabric of reality itself, contributing to advancements in theoretical physics, material sciences, and even biology. These breakthroughs could lead to innovations like new medicines, better climate modeling, and the discovery of new materials with unheard-of properties.

Quantum computing's potential impact on secure communication is another transformative aspect tied to AI. As AI systems increasingly handle sensitive information, the need for robust encryption becomes paramount. Quantum computing introduces quantum cryptography, potentially offering unbreakable security protocols. It leverages the principles of quantum mechanics to create encryption keys that, if tampered with, immediately signal a breach. Such advancements could safeguard personal privacy, financial data, and even national security interests far more effectively than current methods.

However, this transformation does not come without challenges. To capitalize on quantum computing's benefits, we need to rethink existing algorithms and possibly craft entirely new ones—an opportunity ripe for innovation. Compatibility and integration with classical computing systems are clouded with complexities. Significant investments in infrastructure, education, and research are essential to build and maintain quantum computers and their ecosystems.

The business implications are equally profound. Quantum computing could redefine competitive advantages in industries ranging from pharmaceuticals to finance. Imagine a financial institution leveraging quantum computing for dynamic portfolio optimization, making real-time adjustments based on market conditions. Or a pharmaceutical company rapidly discovering new drug formulations by simulating molecular interactions at incredible speeds. This evolution calls for business leaders today to become early adopters, preparing for a future where quantum computing is not just an add-on but a necessity for staying competitive.

One cannot ignore the ethical dimensions either. Quantum AI systems will wield unprecedented power, and with it comes the responsibility to ensure fair and just applications. Safeguards against biases, misuse of power, and the exacerbation of social inequalities must be embedded into the development and deployment processes. Ethical considerations should be integral from the onset rather than afterthoughts.

On the topic of accessibility, the notion of "quantum-as-a-service" could democratize access to quantum computing power. Companies like IBM and Google are already offering cloud-based quantum computing resources, enabling even startups and academic institutions to experiment and innovate without massive capital expenditures. This paradigm shift mirrors the early days of cloud computing, where once prohibitively expensive computational resources became accessible to a broader audience, driving innovation and leveling the playing field.

The educational landscape must evolve in tandem with these technological shifts. Future generations need to be equipped not just with traditional computer science skills but also with an understanding of quantum mechanics and its applications in computing. Curriculums must be reimagined to prepare students for a world where quantum computing and AI are intertwined, fostering a new generation of quantum-literate computer scientists and engineers.

Collaboration will be crucial for unlocking the full potential of quantum computing. International efforts, interdisciplinary projects, and public-private partnerships will drive breakthroughs far more effectively than isolated endeavors. It's akin to building a cathedral; monumental achievements often require the collective effort and expertise of many, working in harmony towards a common vision.

In conclusion, we're standing at the dawn of an era where quantum computing can propel AI into realms we've only dreamed of. It's not merely about faster and more efficient computations; it's about redefining our very approach to solving complex problems and optimizing systems. As quantum computing matures, its synergy with AI will likely herald unprecedented societal and technological advancements, demanding thoughtful consideration and strategic action from tech enthusiasts, business leaders, and policymakers alike.

Advanced Robotics

When imagining the future of AI, thoughts often drift to robots with intuitive intelligence and extraordinary capabilities. This isn't your childhood fantasy of clunky metal humans; we're talking about machines that seamlessly integrate into daily life, significantly enhancing efficiency and quality. Advanced robotics signifies humanity's technological ambitions reaching new heights, blending AI to create machines that can learn, adapt, and remarkably interact with their environment.

What stands out is how these advanced robots aren't just mimicking human actions. They're becoming autonomous entities capable of complex decision-making. Picture a robot in a factory not just following a programmed sequence but dynamically adjusting its actions based on real-time data. The implications for industries and economies worldwide are massive.

Consider the healthcare sector. Advanced robots are already performing delicate surgeries with precision unmatched by human hands. They reduce recovery times, improve outcomes, and make surgeries less invasive. One might wonder, how much further can we push this technology? Well, think exoskeletons, robotic limbs, and even nanobots that swim through our bloodstreams, targeting diseases at the cellular level.

Then there's logistics and supply chain management. Companies are harnessing advanced robots for inventory management, warehousing, and transportation. These robots don't tire, err rarely, and operate 24/7. The efficiency gains are staggering. Automated drones are delivering packages, and self-driving machines are managing warehouses. All of this represents a profound shift towards an AI-driven logistical utopia.

Yet, it's not all smooth sailing. Integration of advanced robotics into the workforce brings about challenges, the most pressing being the displacement of human jobs. However, just as history shows us with previous waves of technological advancement, it's not always about jobs lost but rather jobs transformed. There's potential for humans to work alongside these robots in symbiotic relationships, where robots handle menial tasks, allowing humans to focus on creative and strategic roles.

The manufacturing sector has been a pioneer in adopting robotic technology. Robots capable of assembling intricate electronics or heavy machinery have transformed production lines. Some factories are nearing complete automation. This trend is likely to continue, pushing the boundaries of what's possible in modern manufacturing.

Another compelling aspect of advanced robotics is their role in hazardous environments. Robots can perform tasks too dangerous for humans, such as deep-sea exploration or handling hazardous chemicals. The potential to save lives while accomplishing objectives previously deemed impossible is awe-inspiring. These machines can operate in extreme conditions, from the crushing depths of the ocean to the freezing voids of outer space.

The consumer market is also seeing its share of advanced robotics. Personal assistants, smart home devices, and even robotic pets are becoming parts of everyday life. These innovations make our lives easier, safer, and more enjoyable. Not long ago, having a conversation with a robot or delegating household chores to one seemed like science fiction. Today, it's our reality.

In agriculture, AI-driven robots can monitor crop health, optimize irrigation, and even conduct planting and harvesting. These smart machines aim to address one of humanity's most critical challenges: feeding an ever-growing population. By maximizing efficiency in food production, advanced robotics can significantly contribute to global food security.

Educational institutions are beginning to experiment with robotic aids for teachers. Imagine robots that can provide individualized attention to students, helping those who struggle and challenging those who excel. This could revolutionize the educational landscape, making it more inclusive and personalized.

Of course, we can't discuss advanced robotics without mentioning the ethical considerations. As these machines become more autonomous, questions about accountability, decision-making, and ethical programming arise. How do we ensure these robots make the "right" choices? What safeguards must be put in place to prevent misuse? These are questions that both technologists and ethicists must grapple with as we advance.

The military sector, too, is exploring advanced robotics for tasks ranging from reconnaissance to combat operations. These developments raise significant ethical and political concerns. The idea of autonomous machines making life-and-death decisions is fraught with peril and necessitates strict regulation and international agreements.

A key driver of innovation in advanced robotics is the blend of AI technologies like machine learning and deep learning. These enable robots to continuously improve their performance without human intervention. Imagine a robot chef learning new recipes over time, adapting to culinary trends, and even personal taste preferences. As whimsical as it sounds, the foundational technology is already here.

Strategically, investing in advanced robotics can offer significant competitive advantages. Businesses that adopt these technologies early can leap ahead in productivity and innovation. However, this requires not just investment in technology but also in the workforce. Upskilling and reskilling employees to work alongside advanced robots will be paramount.

In the smart city framework, advanced robots play a crucial role in managing infrastructures such as waste management, road maintenance, and even emergency services. Imagine a scenario where a robotic fleet is responsible for the upkeep of a city, working tirelessly to ensure efficiency and safety. Advanced sensors and AI analytics help these robots make informed decisions on the fly.

The entertainment industry is another area ripe for robotic innovation. The combination of AI and advanced robotics can produce immersive, interactive experiences in theme parks, theaters, and gaming. Imagine animatronics that not only perform preset routines but also interact with the audience, responding to their reactions and creating dynamic narratives.

Innovation in advanced robotics isn't limited to the robots themselves but extends to the entire ecosystem around them. We're talking about advanced materials, new power sources, and innovative interfacing technologies. Collectively, these advancements form a synergistic environment, pushing the boundaries of what's possible.

However, it's crucial to remember that advanced robotics isn't an end but a means to an end—a toolset to help solve complex problems. Whether improving quality of life, optimizing industries, or exploring unknown frontiers, the ultimate goal should focus on making the world a better place for all its inhabitants.

We're at an inflection point. The choices we make today regarding the development and deployment of advanced robotics will shape our society for generations. These decisions shouldn't be taken lightly. Collaborative efforts between technologists, policymakers, businesses, and the public are crucial. We must foster a landscape that promotes responsible innovation, transparency, and inclusiveness.

And so, as we look to the future, we find ourselves in an era rich with potential. Advanced robotics promises to transform our world in ways we are only beginning to fathom. The path forward is filled with challenges, but it's equally brimming with opportunities. By harnessing the collective power of human ingenuity and robotic precision, we can step boldly into this new frontier, ready to redefine the boundaries of our capabilities.

Chapter 25:
The Road Ahead for AI

The road ahead for AI promises to be as thrilling as it is uncertain, with a blend of challenges and boundless opportunities on the horizon. As artificial intelligence continues to break boundaries and redefine possibilities, it's essential for us to adapt and innovate at an unprecedented pace. We're looking at a future where ethical considerations, data privacy, and regulatory frameworks will need to evolve just as rapidly as the technology itself. Potential challenges abound, from addressing algorithmic bias to ensuring equitable access, and the stakes couldn't be higher. Yet, these obstacles are matched by the transformative potential AI holds—from revolutionizing industries and solving global problems to enhancing the quality of life in ways we can't yet fully imagine. As we navigate this landscape, the imperative isn't just to keep up, but to think ahead, strategize wisely, and embrace the endless horizon of possibilities AI ushers in. After all, the future of AI is a journey, one that's just beginning to unfold, and it promises to be an incredible ride.

Potential Challenges

The road to an AI-driven future isn't paved with gold. It's littered with obstacles that could stymie progress in ways both anticipated and unforeseen. One major challenge is the issue of bias. AI systems learn from the data they're fed, and most data isn't inherently neutral. Historical and systemic biases quietly seep into algorithms, and before you know it, you've got an AI that's not just efficient but also discriminatory. Fairness isn't just a noble goal; it's a practical necessity.

Security is another looming mountain to climb. As much as AI promises to bolster cybersecurity measures, it simultaneously offers new avenues for threats. Think of AI systems that can be hijacked to create deepfakes or unleash sophisticated cyberattacks. The stakes are sky-high, and the arms race between AI-driven security solutions and malicious actors never really ends. We're racing against time to secure systems that are complex and always evolving.

Then, there's the issue of data privacy. With AI systems becoming more integrated into our daily lives, the amount of data they process is staggering. How do you ensure that personal data isn't misused or exposed? This challenge is exacerbated in fields like healthcare, where data sensitivity is paramount. Striking a balance between innovation and privacy is a tightrope walk, and the net below is a regulatory framework still catching up with technological advancements.

Not to be neglected is the significant hurdle of regulatory compliance. Governments and institutions are scrambling to put in place rules that can keep up with AI's rapid pace of development. However, there's an inherent tension between stifling innovation and protecting society. Over-regulation can slow down progress, while under-regulation can lead to catastrophic misuse. Navigating this regulatory landscape is akin to steering a ship through stormy seas where the maps are still being drawn.

Moreover, algorithmic transparency is crucial but challenging to achieve. Often, even the developers can't fully explain how an advanced AI makes a particular decision. This "black box" problem creates a trust issue, crucial for applications in critical areas like healthcare, finance, and law enforcement. If people can't understand or trust how decisions are made, public acceptance of AI will face significant resistance.

Workforce displacement is another pressing concern. As AI systems take over repetitive and lower-skilled jobs, millions could find themselves out of work. While there are arguments for job creation through new opportunities, the transition period could be turbulent. Skills training and educational reforms are essential, but implementing these solutions at scale and speed is a monumental task that few countries are fully prepared for.

Interoperability between different AI systems and platforms adds another layer of complexity. In a perfect world, these systems would seamlessly work together, but the reality is that integration can be fraught with incompatibility issues. Different platforms use diverse standards, coding languages, and protocols, making it challenging to create cohesive, unified systems. This not only affects efficiency but can also leave room for security gaps.

Resource constraints present yet another issue. Advanced AI development requires significant computational power, vast amounts of data, and specialized expertise; all resources that are unevenly distributed around the world. This disparity could widen the existing gap between technologically advanced nations and those still developing, leading to a form of digital colonialism where the benefits of AI are concentrated in the hands of a few.

Ethical considerations are profound and multifaceted. Questions about AI's role in decision-making, its impact on human dignity, and its potential to exacerbate inequality are all relevant. For instance, AI's involvement in predictive policing and sentencing could lead to ethical quandaries. If algorithms are prone to bias, how do we ensure justice is served fairly? These moral dilemmas require more than just technical solutions; they call for societal discourse and philosophical introspection.

Moreover, the environmental impact of AI can't be overlooked. Training advanced AI models consumes tremendous amounts of energy, contributing to carbon emissions. As AI becomes more prevalent, its energy footprint becomes a growing concern. Sustainable AI development is not just an option but a necessity for ensuring a balance between technological advancement and planetary health.

Another hurdle revolves around the democratization of AI. While big corporations and financially well-off organizations can afford to invest in cutting-edge AI technologies, smaller businesses and lower-income nations often can't. This creates a monopolistic landscape where a handful of entities hold outsized influence and control over AI advancements. Ensuring that AI benefits can be accessed by a broader audience is not just an issue of fairness, but also of leveraging the full potential of human creativity and problem-solving.

Let's not forget the pace of change itself. Technological shifts are occurring so rapidly that society struggles to keep pace. There's a learning curve involved in understanding AI, its potential, and its pitfalls. The faster AI develops, the harder it becomes for educational systems, businesses, and legal frameworks to adapt. This disparity can lead to a societal divide, separating those who thrive in an AI-driven world from those who struggle to keep up.

Inadequate global collaboration adds another layer of difficulty. While AI has the potential to solve some of humanity's biggest challenges, its benefits are often siloed within national or corporate borders. A more cooperative global approach could unleash tremendous potential in areas like climate modeling, healthcare, and disaster response. Yet, achieving this level of collaboration is easier said than done, given geopolitical tensions and economic rivalries.

Finally, the existential risks posed by super intelligent AI can't be ignored. While it's speculative at this point, the idea that AI could reach a point where it surpasses human intelligence and operates beyond our control is a cause for concern. Whether it's through unintended consequences, misuse, or an outright failure to align AI goals with human values, the potential risks are existential. Addressing these concerns responsibly requires deliberate research, stringent controls, and a robust ethical framework.

Through all these challenges, it's important to recognize that the road ahead isn't simply about identifying obstacles but also about finding solutions. We stand at a pivotal juncture where the decisions we make and the actions we take will shape the future of AI for decades to come. Addressing these challenges head-on, with a spirit of innovation and ethical responsibility, will set the stage for an AI-driven future that benefits all of humanity.

Future Horizons

The horizon for AI gleams with promise, shimmering like a boundless sea of possibilities. It's an exciting time to be involved with technology as we stand on the cusp of achievements that, just a few years ago, seemed like pure science fiction. The future isn't something distant and abstract; it's being sculpted right now in labs, startups, and universities around the world. Let's dive into some of the tantalizing prospects and conceptual dreams that might shape tomorrow.

One of the most thrilling aspects of AI's future is its potential to redefine what we understand as intelligence itself. As AI systems become more sophisticated, they could surpass human capabilities in more nuanced and creative ways. Think about it—imagine AI that can compose symphonies, create works of art, or write novels that stir the soul. While AI has already started making waves in creative fields, the future promises a deeper fusion of human and machine creativity, leading to breathtaking new forms of expression.

Extended into practical applications, AI could elevate industries and processes we haven't even considered yet. Picture agriculture transformed by intelligent systems capable of predicting weather patterns, diagnosing plant diseases early, and even tailoring crop treatments to individual plants. This isn't just about making things more efficient; it's also about potentially breaking new ground in food security and sustainability. AI has the potential to make a bountiful future more attainable, addressing one of humanity's oldest challenges.

Furthermore, the frontier of human health may be redefined by AI. Personalized medicine is just the starting line. Envision a world where AI monitors your health in real-time, preemptively identifies risks before they become serious, and devises custom treatment plans that evolve as you do. We're talking about a healthcare system that moves from a reactive model to a predictive and preventive one. It's about enhancing wellbeing at such an intimate level that it fundamentally changes how we live our lives.

Of course, the societal implications are profound. We're on the brink of potential publications and research breakthroughs that could redefine industries altogether. These advancements promise not only to bolster economies but also to tackle pressing global challenges like climate change and resource scarcity. AI might become an indispensable ally in our quest for a more sustainable future by refining models that predict environmental changes and optimize resource management.

Another thrilling prospect is how AI could transform education. Adaptive learning systems are just the starting point. Imagine learning environments where AI tutors understand each student's unique strengths and weaknesses, crafting personalized learning paths that cater to individual potentials. Geographically bound constraints could dissolve, enabling students from remote and underserved areas to access premier educational resources and mentors. This could democratize education in unprecedented ways.

But what about the existential contemplations? AI ideologies like artificial general intelligence (AGI) and artificial superintelligence (ASI) evoke equal parts excitement and trepidation. AGI, capable of understanding, learning, and applying intelligence across a multitude of tasks, represents a more human-like form of AI. If we cross this threshold, we might unleash a new era where machines collaborate seamlessly with humans in problem-solving and innovation, making AGI partners in our quest for knowledge.

ASI, on the other hand, introduces even more speculative considerations. If AI surpasses human intelligence, the ramifications are both awe-inspiring and puzzling. What new paradigms of existence might this create? How would humanity find its role in a world where machines are the dominant intellects? These questions, while daunting, push us toward uncharted territories in philosophy, ethics, and self-understanding.

The commercial aspect also gleams with potential. AI could become the backbone of entirely new markets and industries. Envision blockchain-powered AI assuring transparent, decentralized decision-making in financial markets or even in democratic processes. The fusion of these technologies could incite revolutionary shifts in society's infrastructure, creating more participatory, transparent, and equitable systems.

Moreover, our cities could transform thanks to intelligent urban planning. Imagine smart cities where AI systems ensure optimal energy usage, manage traffic in real-time, and provide instantaneous emergency responses. These transformations could make urban environments more livable and sustainable, improving quality of life for millions.

And who can overlook the human-machine fusion? Brain-machine interfaces could finally breach the barrier between thoughts and action, enabling paralyzed individuals to move through thought or allowing anyone to interact with devices seamlessly. The line between human and machine may blur, making technologies like augmented reality or cybernetic enhancements a norm rather than an exception.

While these speculations are fascinating, they aren't without challenges. Ethical considerations, data privacy, and regulatory frameworks must evolve alongside these technological advances to ensure AI integrates beneficially into society. As exciting as the future is, it demands responsible stewardship. The road ahead isn't merely about technological prowess; it's also about our collective wisdom to harness and govern that power thoughtfully.

Embracing the future of AI means not just anticipating forthcoming technologies but actively shaping them with principles we hold dear: equity, transparency, and humanity. The narrative of AI isn't merely one of machines and code. It's about harnessing the essence of human ingenuity and steering it toward a future brimming with wonder and possibility.

So, as we peer into the future horizons of AI, we see a tapestry woven with threads of boundless potential, daunting questions, and exhilarating opportunities. It brings to mind the eternal spirit of exploration, the drive to push beyond known limits and uncover what lies beyond. This journey may be as much about rediscovering ourselves as it is about creating new intelligence. Here's to the possibilities that lie ahead, to the convergence of dreams and reality, and to the new dawn on the horizon.

In witnessing the dazzling evolution of AI, one thing's clear: the future will not be built in isolation. It will require collaboration across disciplines, nations, and sectors—an orchestration of diverse talents and perspectives harmonizing towards a common vision. From bustling metropolises to remote villages, everyone has a part to play in this unfolding saga. The future horizons of AI aren't just on the precipice. They're already here, shimmering in the distance, calling us to venture forth.

Conclusion

The journey through the landscape of Artificial Intelligence has been one of discovery, insight, and revelation. From understanding its origins and evolution, delving into the intricate details of its present applications, to envisioning its future impact, we've covered substantial ground. But what does all of this mean for us as tech enthusiasts, business leaders, and even casual observers curious about AI's transformative power?

First and foremost, AI is not just a technological leap; it's a paradigm shift. It challenges established norms and creates new possibilities, often rendering the impossible quite possible. As we stand at the cusp of this AI-driven revolution, it's essential to recognize that AI isn't just about machines learning or algorithms running; it's about a change in perspective.

One of the most compelling aspects of AI is its remarkable potential to solve some of our most pressing global problems. In healthcare, for instance, AI-powered diagnostic tools are already saving lives by identifying diseases at stages when they're most treatable. Personalized medicine, powered by AI, holds the promise of treatments tailored to individual genetic profiles, thereby revolutionizing patient care. Imagine the day when a simple AI prompt could suggest highly effective treatments based on vast amounts of data.

The economic landscape is also poised for transformation. Predictive analytics driven by AI are giving businesses unprecedented insights into market trends, helping companies navigate volatile economic climates with greater confidence. As AI systems become more sophisticated, businesses that harness these capabilities will likely find themselves with a considerable competitive edge, making strategic planning more science than art.

While the benefits are abundant, we must tread carefully regarding personal privacy. AI's hunger for data can lead to significant ethical concerns. It's crucial to establish robust data security measures and ethical guidelines to ensure that personal information is not exploited. Balancing technological advancements with ethical considerations will be key as we move forward.

In the business realm, AI is enhancing operational efficiency in ways that were unimaginable a decade ago. From streamlining supply chains to personalizing customer experiences, the integration of AI systems is redefining business models. The future will favor organizations that can adapt and integrate AI strategies into their core operations, making resilience and adaptability vital skills for future business leaders.

However, the workforce faces a transitional challenge. While AI might displace certain jobs, it will also create new opportunities. Embracing this change involves learning new skills and being open to evolving career paths. It's about preparing for a future where continuous learning and adaptability become the norm, ensuring that the human workforce remains relevant and valued.

In transportation, AI is paving the way for autonomous vehicles and smart traffic systems, promising safer and more efficient travel. Imagine a world where traffic jams are relics of the past, replaced by smoothly flowing lanes governed by intelligent systems that optimize traffic flow in real-time. The ripple effects on urban planning, commuting, and even environmental sustainability are profound.

Communication technologies are set to become more intuitive, thanks to AI advancements in natural language processing and voice assistants. The future of human-machine interaction looks increasingly conversational and personalized, breaking down barriers to access and usability.

The entertainment industry is also experiencing a renaissance with AI. Personalized recommendations are becoming increasingly sophisticated, ensuring that content reaches audiences more effectively. Moreover, AI is playing a significant role in content creation, from generating realistic graphics to assisting in scriptwriting, thereby revolutionizing the production landscape.

When we look at education, the promise of AI becomes even more exciting. Adaptive learning systems and virtual tutors can personalize education to each student's needs, creating a more engaging and effective learning environment. This could democratize education, making high-quality learning accessible to people around the globe, irrespective of socioeconomic status.

In environmental sustainability, AI's role is paramount. Climate modeling powered by AI can provide more accurate predictions, aiding in our fight against climate change. Resource management systems can ensure that we use our planet's resources more efficiently and sustainably, providing hope for a greener future.

But with such power comes the necessity for robust security. AI's integration into cybersecurity and national defense systems holds the potential to ward off growing cyber threats and ensure national security in a digital age. As AI becomes more integrated into these areas, its role in safeguarding our societies will become increasingly indispensable.

However, AI's societal impacts can't be ignored. The challenge of social inequality could be exacerbated by AI if not managed properly.

Public perception of AI, driven by media and individual experiences, will also play a crucial role in its adoption and regulation. Society must navigate these waters carefully, ensuring that AI serves as a tool for empowerment rather than exclusion.

AI's stimulation of innovation can't be overstated. From enhancing research and development to streamlining patent analysis, AI accelerates the pace of discovery and implementation. It allows for smarter, more efficient processes that push the boundaries of what we thought was possible.

Real-world applications of AI, as explored through various case studies and industry implementations, provide a practical lens through which to view its capabilities. These examples serve as testament to the transformative power of AI, offering a blueprint for others to follow.

Ethical considerations in AI development remain critical. We must continually address biases, ensure accountability, and strive for fairness in algorithms to foster trust and widespread adoption. Transparent and inclusive practices can pave the way for a more ethical AI future.

The regulatory landscape will inevitably evolve. Existing laws need to be revisited, and new policies must be crafted to address the unique challenges posed by AI. Engaging in proactive policy-making can help navigate these challenges, ensuring that progress is coupled with responsibility.

Investing in AI, both financially and in terms of skill development, is essential. Market analysis and venture capital investments signal the growing importance of AI in the global economy. Ensuring that we are prepared for this AI revolution involves strategic planning, skill development, and a commitment to lifelong learning.

Expert opinions offer invaluable insights, guiding us through predictive analyses and shaping our understanding of AI's trajectory.

Listening to voices from various fields enriches our perspective and prepares us for the multifaceted journey ahead.

Global perspectives on AI reveal a tapestry of diverse approaches and collaborative efforts. International collaboration can drive advancements and ensure that AI benefits humanity as a whole, rather than a select few. Embracing global thinking will be key in realizing the full potential of AI.

Technologies like quantum computing and advanced robotics, when combined with AI, hint at a future where the boundaries of innovation are continually pushed. These future technologies promise to unlock new dimensions of capability and performance, expanding what AI can achieve.

The road ahead for AI is paved with both challenges and remarkable potential. Future horizons suggest that we have only scratched the surface of what AI can accomplish. However, potential challenges, such as unintended consequences and systemic biases, must be managed with foresight and care.

In conclusion, the profound impact of AI on various facets of society is undeniable. By understanding AI, embracing its potential, addressing its challenges, and guided by ethical considerations, we are better equipped to navigate the AI-driven future. It's an exciting time, filled with opportunities and responsibilities, and the path we choose will shape the world for generations

Appendix A:
Appendix

As we wrap up our exploration into the transformative realm of artificial intelligence, this appendix serves as your go-to section for clarifying concepts, revisiting key ideas, and gathering additional resources for deeper learning. It's designed to complement everything we've covered and act as a handy resource you can come back to at any time.

Key Definitions and Terms

Throughout the book, we've tackled a wide array of technical jargon and industry-specific terms. Here's a curated list of the most crucial terms to keep you on solid footing:

Artificial Intelligence (AI): The branch of computer science focused on building smart machines capable of performing tasks that typically require human intelligence.

Machine Learning (ML): A subset of AI that involves training algorithms to make predictions or decisions based on data.

Deep Learning: A specialized form of ML using neural networks with many layers, enabling more complex data processing.

Natural Language Processing (NLP): The technology enabling machines to understand, interpret, and respond to human language.

Autonomous Vehicles: Self-driving cars and other vehicles that use AI technologies to navigate without human intervention.

Additional Resources

Knowledge is power, especially when venturing into the vast universe of AI. The following resources are excellent for deep dives into specific topics we've touched on:

Books: Look for titles by experts like Stuart Russell and Peter Norvig. They provide solid foundations and nuanced insights into AI principles and applications.

Online Courses: Platforms such as Coursera, edX, and Udacity offer beginner to advanced courses led by industry giants like Andrew Ng and Sebastian Thrun.

Research Papers: Websites like arXiv.org host a treasure trove of AI research papers, delving into everything from theoretical frameworks to cutting-edge innovations.

Tools and Platforms

If you're eager to get hands-on with AI, numerous tools and platforms can kickstart your journey:

TensorFlow: Google's open-source library for machine learning and deep learning applications.

PyTorch: Another popular deep learning library, developed by Facebook's AI Research lab.

OpenAI Gym: A toolkit for developing and comparing reinforcement learning algorithms.

Final Thoughts

The world of AI is as challenging as it is exhilarating. The knowledge and strategies we've discussed are stepping stones on your journey into this transformative field. Whether you're a business leader eyeing innovative solutions, a tech enthusiast curious about the latest trends, or simply someone captivated by the future's potential, remember that the adventure is just beginning. The tools, concepts, and insights we've explored are gateways to a future where AI serves as a catalyst for unprecedented progress and creativity.

As we move forward, keep questioning, stay curious, and embrace the dynamic landscape of AI. The road ahead is filled with boundless opportunities waiting to be discovered.

Glossary of Terms

Welcome to the glossary, your quick reference guide to some of the key terms and concepts discussed throughout the book. Think of this section as your trusty companion while navigating the intriguing world of Artificial Intelligence. Let's dig in!

Adaptive Learning Systems - Intelligent systems that tailor educational content to meet the unique learning needs of individual students.

Advanced Robotics - Robots equipped with AI capabilities, enabling them to perform complex tasks with high levels of autonomy.

Artificial Intelligence (AI) - The simulation of human intelligence in machines programmed to think and learn.

Autonomous Vehicles - Self-driving cars and other transport modes that use AI to navigate and make decisions without human intervention.

Bias and Fairness - Ethical considerations around ensuring AI systems do not discriminate or act unfairly towards any group.

Case Studies - Real-world examples of AI implementation, showcasing successes, challenges, and learnings.

Climate Modeling - The use of AI to predict future climate patterns based on historical data and complex algorithms.

Content Creation - The use of AI to generate music, art, written text, and other creative works.

Cybersecurity - The practice of protecting systems, networks, and data from digital attacks, with AI playing a crucial role in threat detection and response.

Data Security - Ensuring that data is protected from unauthorized access and breaches, particularly critical in the AI era.

Deep Learning - A subset of machine learning involving neural networks with many layers, enabling high-level data abstractions.

Diagnostic Tools - AI applications that assist in the detection and diagnosis of health conditions.

Machine Learning - A method of data analysis that automates analytical model building, allowing computers to learn from data and improve their performance over time.

Market Trends - Predictive insights driven by AI to understand and forecast the ever-changing dynamics in various markets.

Natural Language Processing (NLP) - A field of AI that focuses on the interaction between computers and humans through natural language.

Operational Efficiency - Improvements in business processes and workflows facilitated by AI to increase productivity and reduce costs.

Patent Analysis - The use of AI to sift through and analyze patents, identifying trends and innovation opportunities.

Personalized Medicine - Tailoring medical treatment to the individual characteristics of each patient, guided by AI analysis.

Personalized Recommendations - AI-driven suggestions tailored to individual preferences, commonly used in e-commerce and entertainment platforms.

Predictive Analytics - The use of statistical algorithms and AI to make predictions about future events based on historical data.

Quantum Computing - The next frontier in computing, promising exponential increases in speed and capacity, and heavily interlinked with the future of AI.

Resource Management - AI applications designed to optimize the use and allocation of resources, from energy to raw materials.

Smart Traffic Systems - AI-driven technologies aimed at optimizing traffic flow and reducing congestion.

Virtual Tutors - AI-driven educational tools that provide personalized instruction and support to students.

Voice Assistants - AI-powered applications like Siri and Alexa that interact with users via voice commands.

Utilize this glossary as your quick-reference toolkit to enhance your understanding of AI and its multifaceted impact on our world. Stay curious, stay informed, and embrace the future with confidence!

Further Reading

We've journeyed through a vast expanse of AI concepts, dissecting its history, exploring present trends, and peering into its future implications. But the world of artificial intelligence is vast, constantly evolving, and brimming with untold facets. For those eager to dive deeper, there's a wealth of literature that stretches beyond the pages of this book. Whether you're a tech enthusiast craving the latest in AI advanceements or a business leader seeking strategies to harness its power, the following resources will enhance your understanding and sharpen your insights.

First, consider delving into the foundational texts that have shaped the AI landscape. Books such as "Artificial Intelligence: A Modern Approach" by Stuart Russell and Peter Norvig offer a comprehensive look at the principles and techniques that form the backbone of AI.

They're often regarded as essential reading for anyone looking to grasp the theoretical underpinnings and practical applications of the field.

For a more historical perspective, "Machines Who Think" by Pamela McCorduck provides a captivating journey through the development of AI, touching on the pioneers and pivotal moments that have contributed to today's cutting-edge innovations. And then there's "The Master Algorithm" by Pedro Domingos, which peels back the layers of machine learning to reveal the algorithms driving progress and the quest for a unified theory of AI.

Staying updated with the latest research and development is crucial. Journals like "Journal of Artificial Intelligence Research (JAIR)" and "Machine Learning Journal" provide peer-reviewed articles on the newest discoveries, experiments, and theoretical advances. Additionally, arXiv.org is an excellent repository of pre-publication papers and research findings, covering a breadth of topics within the AI domain.

If you're particularly interested in the technical aspects and emerging technologies in AI, books like "Deep Learning" by Ian Goodfellow, Yoshua Bengio, and Aaron Courville present a thorough elucidation on neural networks and deep learning. These texts not only discuss the theory but also offer practical insights and applications, making them invaluable for both academics and practitioners alike.

Shifting focus to the societal impacts of AI, "Weapons of Math Destruction" by Cathy O'Neil showcases the darker side of data science and algorithms, revealing how they can perpetuate inequality and create systemic problems. Meanwhile, "AI Superpowers" by Kai-Fu Lee compares the AI ecosystems of China and the United States, providing a geopolitical lens on the future race for AI supremacy.

Business leaders might find "Prediction Machines: The Simple Economics of Artificial Intelligence" by Ajay Agrawal, Joshua Gans, and Avi Goldfarb particularly enlightening. This book breaks down how AI can be leveraged to enhance decision-making processes and create value within organizations. Combined with "Human + Machine: Reimagining Work in the Age of AI" by Paul R. Daugherty and H. James Wilson, it provides a robust guide to integrating AI seamlessly into business operations.

Tech enthusiasts interested in the ethical considerations and philosophical questions around AI should look into "Super-intelligence: Paths, Dangers, Strategies" by Nick Bostrom. It delves into the potential existential risks of AI and contemplates how to navigate the path to superintelligent systems safely. Similarly, "Life 3.0: Being Human in the Age of Artificial Intelligence" by Max Tegmark ponders the long-term future of AI and its implications for humanity.

To stay abreast of current developments, active participation in communities and forums like AI Alignment Forum, GitHub discussions, and Stack Overflow can be beneficial. These platforms offer not only the latest code and project insights but also lively debates and collaborative opportunities with AI enthusiasts and experts worldwide.

Podcasts and webinars provide a more conversational and often more digestible source of information. Programs like "The AI Alignment Podcast" or "Data Skeptic" frequently host leading thinkers who distill complex ideas into accessible discussions, offering fresh perspectives and real-world insights. TED Talks on AI also showcase thought leaders and innovators sharing their visions and breakthroughs in bite-sized, engaging formats.

For a more policy-oriented approach, consider reports and white papers from think tanks and research institutions like the Brookings Institution, the AI Now Institute, and the Future of Life Institute. These documents often provide in-depth analyses of the implications of AI, recommendations for guidelines, and strategic insights into AI governance.

Last but not least, involvement in professional associations such as the Association for the Advancement of Artificial Intelligence (AAAI) or the Institute of Electrical and Electronics Engineers (IEEE) can be immensely valuable. These organizations frequently host conferences, workshops, and seminars where the cutting edge of AI research and applications are discussed among experts and practitioners.

As we look to the future, it's clear that continuing education and proactive engagement with AI literature and community are paramount. The recommended readings above offer a spectrum of knowledge, from the nuanced technical intricacies to the broader societal ramifications, ensuring you're well-equipped to both understand and influence the AI-driven world.

In sum, AI's dynamic landscape requires a commitment to continuous learning. With these resources, you'll not only stay ahead of the curve but also contribute thought leadership in your respective field. Embrace this journey of discovery, armed with the tools and knowledge to turn curiosity into expertise and vision into reality.